NATURAL HAZARDS ATLAS OF JAMAICA

Jamaica is susceptible to four main natural hazards: floods, landslides, earthquakes and hurricanes.

NATURAL HAZARDS ATLAS OF
JAMAICA

PARRIS LYEW-AYEE JR AND RAFI AHMAD

Mona GeoInformatics Institute
University of the West Indies, Mona

University of the West Indies Press
Jamaica • Barbados • Trinidad and Tobago

University of the West Indies Press
7A Gibraltar Hall Road, Mona
Kingston 7, Jamaica
www.uwipress.com

Mona GeoInformatics Institute
University of the West Indies, Mona
Kingston 7, Jamaica
www.monagis.com

A catalogue record of this book is available from the National Library of Jamaica.

ISBN: 978-976-640-259-4

Book design by Robert Harris.
Cover design by Dominique Thompson and Robert Harris.
Set in Plantin 11/16 x 36.
Printed in China by Regent Publishing Services.

PHOTO CREDITS

Rafi Ahmad, University of the West Indies
Luke Buchanan, Mona GeoInformatics Institute
Alexander Grennell, Mona GeoInformatics Institute
Christina Francis, Mona GeoInformatics Institute
Parris Lyew-Ayee Jr, Mona GeoInformatics Institute
Water Resources Authority (WRA)
Jamaica Information Service Photo Library (JIS)
Institute of Jamaica (IOJ)
The *Gleaner* Archives, Jamaica

DISCLAIMER

This atlas uses information gleaned from many different sources over many years of research, including field and library archive research, published papers and models, and from various government agencies. Hazard inventories are collected from historical media archives, themselves not official scientific reports. However, these archives go back (in some cases, over 170 years) well before any dedicated scientific studies on hazards began in Jamaica. The atlas therefore covers a greater time scale of hazard coverage, and also becomes a useful tool to place more recent hazard models into a historical and/or regional context. All data have been fully adapted for inclusion in the atlas. All data used have been authorized by their owners if not originally created by the Mona GeoInformatics Institute or the Unit for Disaster Studies. The atlas is not intended to be an official record, but merely a source for general contextual information on hazards.

CONTENTS

ILLUSTRATIONS

PLATES

TABLES

ABOUT THE ATLAS

Natural hazards are the results of natural processes that continually shape the physical environment over time. They occur at many different time scales and range from slow and gradual processes, such as erosion and deposition, to sudden and catastrophic events, such as earthquakes and hurricanes, which, in an instant, can dramatically alter the landscape.

Jamaica is susceptible to four main natural hazards: floods, landslides, earthquakes and hurricanes. Over its recorded history, Jamaica has had numerous disaster events resulting from the impacts of any of these hazards. In recent years, there have been increasing efforts to integrate hazard mitigation into national planning activities, as well as into public education and awareness. This atlas is meant to present existing research and information in a jargon-free, easy-to-use manner for the general public.

The atlas is divided into three separate sections. **Section 1** looks at the physical geography of Jamaica: its landscape and landforms, geology, and natural environment (rivers and waterways, protected areas and the like). This includes custom maps created specifically for this atlas. Data displayed on these maps originate from multiple sources (such as the Mines and Geology Division, the Water Resources Authority and the National Environment and Planning Agency). **Section 2** illustrates the four different hazards that commonly affect Jamaica: earthquakes, hurricanes, floods and landslides. Each event is accompanied by maps and models showing historical impacts and modelled scenarios. This section draws primarily from research at the Mona GeoInformatics Institute and the Unit for Disaster Studies. **Section 3** has maps of each parish of Jamaica and shows larger-scale maps and models of each hazard at the local level. Events occurring within each parish are illustrated in greater detail, accompanied by the date of that event, as well as explanations for the prevalence of certain events in particular locations. Individual parish summaries accompany each subsection, along with a socio-economic summary followed by its unique natural hazard profile. Each section has the contemporary road network and settlement distribution, which provides a point of reference for each map and juxtaposes the physical and social environments. Photographs, graphs and charts, as well as annotations and helpful statistics, are included throughout.

ACKNOWLEDGEMENTS

We would like to thank the following individuals for their valuable assistance in the preparation of this atlas. We greatly appreciate the contributions of the staff at the Mona GeoInformatics Institute, namely projects manager Karen McIntyre, senior field officer Luke Buchanan, deputy director Dr Ava Maxam, assistant projects manager Helen Liu, and GIS officer Christina Francis. We also greatly acknowledge the assistance of Anne Lyew-Ayee of the Department of Geography and Geology in the content editing of the final draft.

We would also like to recognize the help from the staff of the University of the West Indies Press for their strong support of this project, particularly director Linda Speth.

Finally, we would like to recognize the strong support given to this project from members of the insurance sector, particularly NEM Insurance, and individuals from both central and local governments, in particular, the Water Resources Authority.

We hope that this atlas serves you well.

DATA SOURCES

Maps

Geology map layer adapted from Alan Fincham, 1977. Cartography and lithography by the Survey Department, Jamaica; Geological Data prepared by the Mines and Geology Division, Ministry of Mining and Natural Resources, Jamaica; originally compiled by N. McFarlane (1977); Drafted by: D. Lyew-Ayee (1977); first edition edited and corrected by S. Brookes (1948).

Protected areas digitized from a combination of maps obtained from the Jamaica Protected Areas Trust Limited, 2009 and the National Environment and Planning Agency, 2005.

Thirty-year mean rainfall derived from the Meteorological Service of Jamaica's listing of meteorological station data and their coordinates; subsequently plotted and interpolated to raster model by the Mona GeoInformatics Institute.

Hydrological basins across Jamaica: Spatial data obtained from the Water Resources Authority.

Hydrostatigraphy map of Jamaica: Spatial data obtained from Water Resources Authority.

Hurricane pathways: Data obtained online from National Oceanic and Atmospheric Administration; subsequently plotted by Mona GeoInformatics Institute. http://www.csc.noaa.gov/hurricanes/#.

Wind speed 10-, 25-, 50- 100-year return period: maps obtained from the Caribbean Disaster Mitigation Project, 2009. http://www.oas.org/CDMP/document/reglstrm/index.htm

Landslide engineering classes and recommendations adapted from the Geotechnical Engineering Office, *Guide to Rock and Soil Descriptions*, 5th ed. (Hong Kong: Geotechnical Control Office, Hong Kong, 2000). http://www.cedd.gov.hk/eng/publications/geo/doc/eg3_checklist.pdf

Modified Mercalli intensity table: Bruce A. Bolt, *Abridged Modified Mercalli Intensity Scale, Earthquakes: Newly Revised and Expanded* (New York: W.H. Freeman and Co., 1993), appendix C.

Parish Maps (Map 3.2.1–3.14.2)

Landslide and flood events sourced from Mona GeoInformatics Institute fieldwork as well as *Gleaner* archives. Reports converted to locational data by the Mona GeoInformatics Institute.

Contours digitized from the 1:12,500, Jamaica Topographic Map Series, Survey Department, National Land Agency.

Hotel, churches, hospital, police stations, points of interest mapped by means of Mona GeoInformatics Institute fieldwork (JAMNAV).

Buildings digitized by the Mona GeoInformatics Institute from IKONOS 2001 satellite imagery.

Trigonometrical stations and urban and industrial areas polygon digitized by the Mona GeoInformatics Institute from the 1:50,000 Jamaica Topographic Map Series, Survey Department, National Land Agency.

All Maps

Roads layers and names obtained by Mona GeoInformatics Institute field collection (JAMNAV) as well as digitization from IKONOS satellite imagery.

Rivers and parish capital polygons digitized by the Mona GeoInformatics Institute from the 1:50,000 Jamaica Topographic Map Series, Survey Department, National Land Agency.

Digital elevation model created by the Mona GeoInformatics Institute by spatial interpolation of contours digitized from 1:12,500 Jamaica Topographic Map Series, Survey Department, National Land Agency.

INTRODUCTION

Natural hazards threaten countries in many different ways: they affect individuals, public and private properties, livelihoods, economies, a country's development, even the stability of governments. The distribution of, and therefore exposure to, hazards is a function of many factors. Development may exacerbate or mitigate hazard impacts and choices of settlement location, design and layout may influence a hazard's impact. However, hazard impacts are inevitable; how individuals and societies prepare for these depends on the level of information made available to them.

There are many modes of promoting hazards awareness in Jamaica. Many websites exist for the general public to access, and many dedicated agencies facilitate hazard awareness and mitigation at regional, national and community levels. Hazard vulnerability is considered in every request for development approval and insurance coverage. But, in general, the hazard information available locally is generic. For example, following significant rainfall events, the public is warned of likely flooding in the low-lying areas and landslides, yet the low-lying areas prone to flooding and landslide are not explicitly mentioned. Roads are blocked by landslides and are subsequently cleared, but the frequency of landslides, which would help predetermine the areas more prone to landslides in a particular event, is often ignored. Although local knowledge is used in community hazards preparedness, large-scale planning is done at the level of the parish council or the central government, and those parties may use more general data as the basis of their recommendations.

Information to guide avoidance of hazard impact in the local context is critical. However, the public at large may be only nominally aware of hazards, and may not associate this knowledge with their own community and its specific risks. Moreover, developers ignorant of local hazard threats endanger their investments, as well as lives and properties.

Natural Hazards Atlas of Jamaica is designed to present readily available information in a graphic and stylized format that is accessible to a general audience while being of value to decision-makers in local and central governments, developers, community stakeholders, businesses, schools and universities, and international donor agencies. The atlas attempts to connect these disparate groups around common and basic information. It is not intended to be a definitive textbook on the subject; rather, it is aimed at promoting hazards awareness across society in a relevant manner. An atlas, rather than a map book, presents useful facts and figures associated with an event or location, which, along with pictures and graphs, provide ready information to readers. By looking at real events, both recent and ancient, readers are taken on a journey across Jamaica, through time and space, where hazards rarely impact in isolation.

Facts and Figures

JAMAICA'S COUNTRY PROFILE

1. Land area: 10,992 square kilometres

2. Coastline length: 1,022-kilometre continental shelf 200NM or to edge of the continental margin (*Economic and Social Survey of Jamaica*, 2008)

3. Population: 2.8 million – urban: 1.7 million; rural: 1.116 million (*CIA World Fact Book*, 2010)

4. Population density: 223 persons per square kilometre (*Economic and Social Survey of Jamaica*, 2008)

5. National growth rate: 0.85 per cent per annum, 1950–2003 (*Economic and Social Survey of Jamaica*, 2008)

6. National gross domestic product: US$1,047,521 million, 2008 estimate (*Economic and Social Survey of Jamaica*, 2008)

7. National gross domestic product per capita: US$8,400 (*CIA World Fact Book*, 2010)

8. Public debt: 106.9 per cent of gross domestic product (*Economic and Social Survey of Jamaica*, 2008)

9. External debt: US$6,343.72 million (*Economic and Social Survey of Jamaica*, 2008)

10. Population below poverty line: 14.8 per cent (*CIA World Fact Book*, 2010)

– SECTION 1 –

PHYSICAL ENVIRONMENT

Facts and Figures

PHYSICAL ENVIRONMENT

1. Jamaica may be described according to five general landscapes: coastal terrain; inland plains; river valleys; limestone hills and plateau; and volcaniclastic mountainous terrain.

2. The geology of Jamaica may be broadly described according to limestones, volcaniclastics and alluvial deposits.

3. There are 10 declared or proposed protected areas in Jamaica.

4. Rio Minho is the longest river in Jamaica.

5. Black River is the longest navigable river in Jamaica.

6. Rio Grande has the highest average flow ($25m^3/sec$).

7. Limestone aquifers make up 50% of Jamaica.

8. Alluvium (sand and gravel) aquifers make up 8% of Jamaica.

9. Water supply is presently 84% by volume from aquifers (80% limestone and 4% alluvium) and 16% from surface water.

10. Water resources reserves are 90% groundwater in limestone aquifers.

1.1 LANDSCAPE

Disasters are related to the landscapes in which they occur, and each landscape is characterized by its unique susceptibility to hazards. A tsunami, for example, is not expected to impact inland, mountainous areas; a landslide is unlikely to occur on flat land. Jamaica may be described according to five general landscapes:

- coastal terrain
- inland plains
- river valleys
- limestone hills and plateaux
- volcaniclastic mountainous terrain

Each landscape is a product of different geological, surficial and meteorological processes. Landscapes may be shaped by man-made activities such as urbanization, farming, mining and recreation.

Coastal terrain describes low-lying regions located along the shore. For the purposes of this atlas, coastal terrain is defined as areas at or below 10 metres above sea level; it is more a function of elevation than distance from the shore. These low-lying areas are subject to coastal erosional and depositional processes, and they are vulnerable to coastal hazards such as storm surges (associated with tropical storms and hurricanes) and tsunamis (associated with earthquakes).

Inland plains have largely gentle topographies, although they may not necessarily be low-lying (as measured by elevation above sea level). Inland plains result from alluvial deposition from rivers flowing through or from lateral erosion in karst interior valleys. Inland plains are vulnerable to flooding from the overflow of rivers or from surface run-off.

River valleys are the products of dissection of (usually) volcaniclastic hillslopes. They may be the scenes of flooding (from both water and fluidized debris) and landslides. River valleys may have permanent stream flow or ephemeral streams carrying water only after heavy or prolonged rainfall.

Limestone hills and plateaux occupy over 50 per cent of Jamaica's land area. They are characterized by the lack of surface water as well as, in some places, bauxite in-fill. Karst flooding is common in areas as well, when the water table is near the surface. These areas are significant sources of groundwater. Limestone plateaux, such as the Manchester Plateau, also reach relatively high altitudes and are subjected to intense winds.

Volcaniclastic mountainous terrain describes heavily dissected landscapes, including the Blue Mountains. Such terrain includes igneous, metamorphic and sedimentary formations, and is vulnerable to landslides and flooding. The regions are typically intensely faulted and are associated with many earthquakes affecting Jamaica. These areas are also high-altitude regions and have high exposure to extreme winds. Volcaniclastic mountainous terrain is associated with river valley landscapes, although the latter have more active processes at play and warrant a separate classification.

Table 1.1.1 Number of Buildings on Each Type of Landscape Across Jamaica

Landscape	Number of Buildings
Coastal	61,736
Inland plain	152,739
Limestone upland	185,660
Non-limestone upland	14,455
River valley	90,223

Source: Mona GeoInformatics Institute (MGI) buildings and landscape analysis, 2009.

Plate 1.1.1 Aerial view of Kingston (A. Grennell, 2008)

Plate 1.1.2 Bog Walk Gorge (R. Ahmad, 2006)

Plate 1.1.3 Coastal erosion, White Horses and Rozelle, St Thomas (R. Ahmad, 2011)

Plate 1.1.4 Cockpit Country (P. Lyew-Ayee Jr, 2003)

Plate 1.1.5 Blue Mountain Peak (P. Lyew-Ayee Jr, 2011)

18°30'N

IRONSHORE
BARRETT TOWN
Falmouth
Montego Bay
DISCOVERY BAY
RUNAWAY BAY
St. Ann's Ba
MOUNT SALEM
DUNCANS
Lucea
HOPEWELL
ADELPHI
OCH
HANOVER
ANCHOVY
WAKEFIELD
CLARKS TOWN
ST. JAMES
BROWNS TOWN
BAMBOO
STEWART TOWN
TRELAWNY
CLAREMONT
GOLDEN GR
MAROON TOWN
RAMBLE
ULSTER SPRING
ALEXANDRIA
CAMBRIDGE
GRANGE HILL
ST. ANN
CATADUPA
ALBERT TOWN
NEGRIL
MONE
WESTMORELAND
WAIT-A-BIT
PETERSFIELD
LITTLE LONDON
DARLISTON
ACCOMPONG
GINGER HILL
CHRISTIANA
LLUIDAS V
Savanna-la-Mar
BALACLAVA
MAGGOTTY
MILE GULLY
MANCHESTER
CHAPELTON
MIDDLE QUARTERS
LACOVIA
CLARENDON
WILLIAMSFIELD
SANTA CRUZ
Black River
ST. ELIZABETH
Mandeville
May Pen
18°N
MALVERN
NEWPORT
LITITZ
OL
JUNCTION
CROSS KEYS
BULL SAVANNA
ALLIGATOR POND
LIONEL TOWN

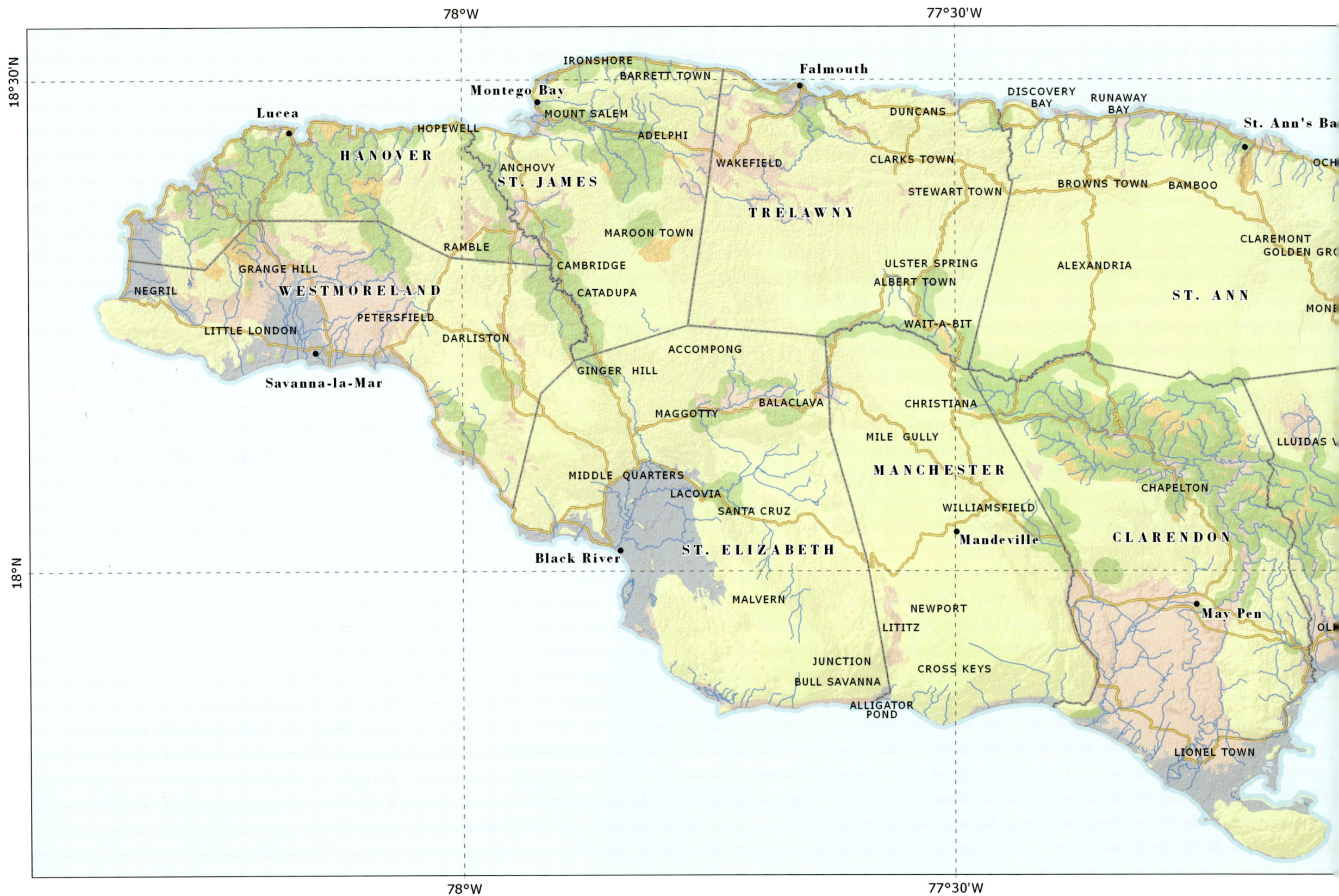

Map 1.1.1 Landscape types across Jamaica

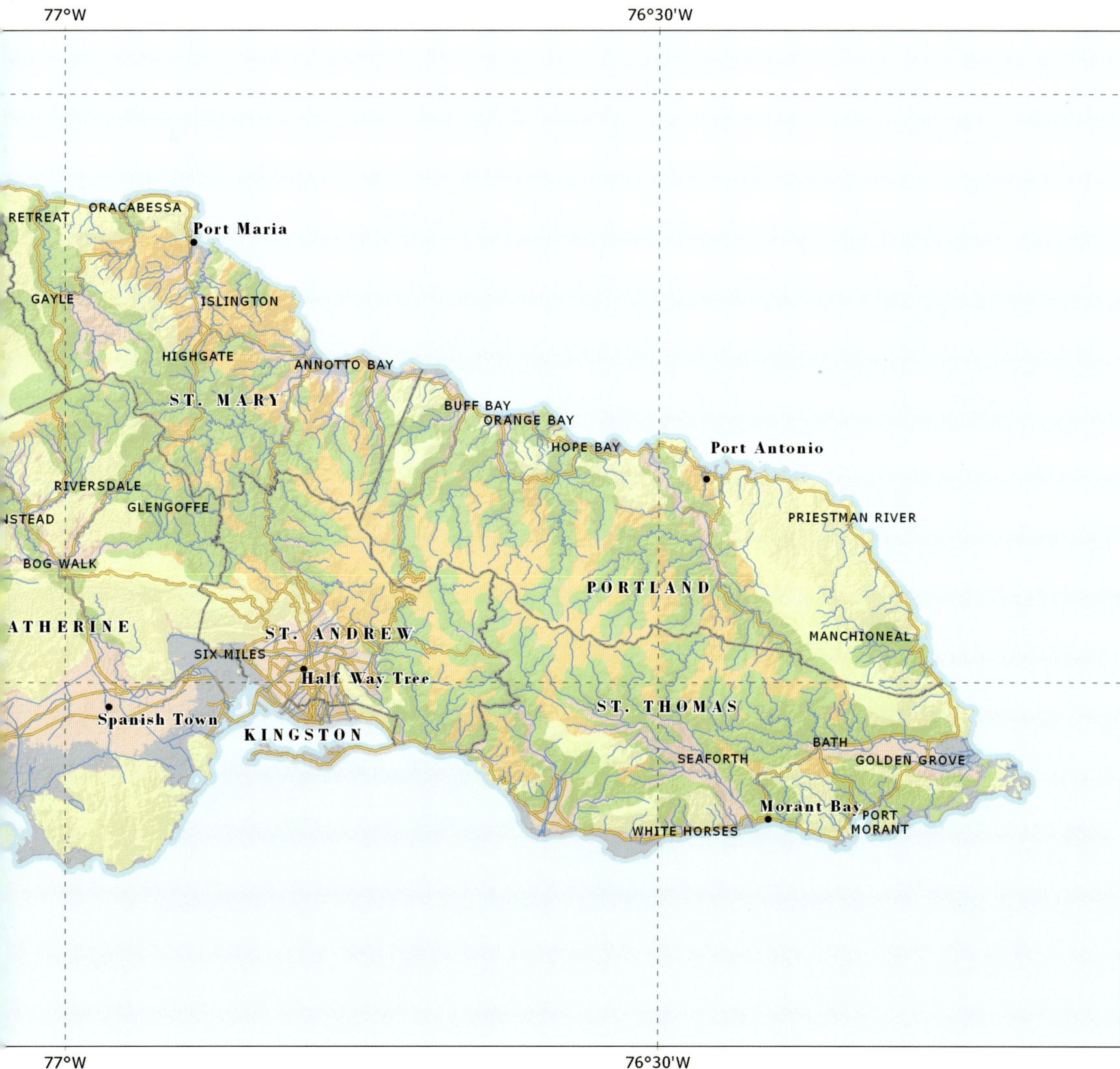

77°W 76°30'W

18°30'N

Landscape

Limestone hill and plateau

Volcaniclastic mountainous terrain

Inland plain

River valley

Coastal terrain

● Parish capital

Parish boundary

Main road

Major river

RETREAT ORACABESSA
Port Maria
GAYLE ISLINGTON
HIGHGATE ANNOTTO BAY
ST. MARY BUFF BAY
ORANGE BAY
HOPE BAY Port Antonio
RIVERSDALE PRIESTMAN RIVER
GLENGOFFE
NSTEAD
BOG WALK PORTLAND
ATHERINE ST. ANDREW MANCHIONEAL
SIX MILES
Half Way Tree
Spanish Town ST. THOMAS
KINGSTON BATH
SEAFORTH GOLDEN GROVE
Morant Bay
WHITE HORSES PORT
MORANT

18°N

1:450,000

0   5   10   20 kilometres

0   5   10   20 miles

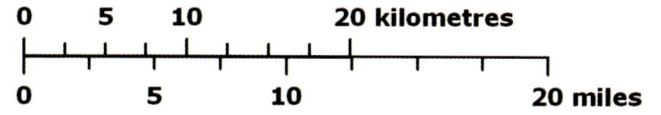

Map Datum:
Latitude-Longitude geodetic grid
World Geodetic System 1984 datum
Prime Meridian: Greenwich
Angular unit: Degree

77°W 76°30'W

1.2 GEOLOGY

The nature of the bedrock and geological structure define the landforms of Jamaica. The bedrock in Jamaica is highly fractured. Major and minor faults criss-cross the island and these fractures serve as lines of weakness in the landscape, either as conduits for weathering or associated with earthquakes and tremors. Jamaica's geology may be broadly described according to the following groups.

- **Limestones** – Jamaica is divided into two broad groups: the pure White Limestone Group and the impure Yellow Limestone Group. The former group is associated with significant karst development, and includes the regions with greatest aquifer development. Most springs in Jamaica originate in these areas.

- **Volcaniclastics** – These are made up of igneous, sedimentary and metamorphic rocks, and comprise the oldest rocks in Jamaica. Major surface drainage occurs in areas underlain by volcaniclastic bedrock.

- **Alluvial deposits** – These are largely fluvial deposits, laid down by rivers over many millennia. Alluvial deposits are largely developed along the southern plains of Jamaica, and they are the sites of large-scale plantation cultivation.

Caves occur dominantly in the limestone areas of Jamaica. Sculpted by the movement of water, their formation may be assisted by the presence of joints and fault lines, which are widened by water movements. Caves also represent subsurface solutional processes which continually occur. Construction may be at risk of subsidence if built on a cave network: cave systems can transport underground waste vast distances, discharging polluted water as springs, and reactivation of former underground waterways during heavy rainfall periods can lead to flooding.

Note: Many infrastructural references made in this atlas with respect to different hazards typically involve police stations, fire stations and hospitals as emergency services, and official government shelters. Their individual and collective exposure to hazard threats have a greater potential impact than other non-critical facilities.

Table 1.2.1 Number of Buildings in Relation to Major Fault Zones

Total	Police Stations	Fire Stations	Hospitals	Shelters
Number of buildings	1,050	184	33	35
Number of buildings within 100 metres of fault	110	18	1	2

Source: Mona GeoInformatics Institute (MGI) JAMNAV data locations and analysis, 2009.

Plate 1.2.1 Guys Hill Formation, Clarendon (P. Lyew-Ayee Jr, 2011)

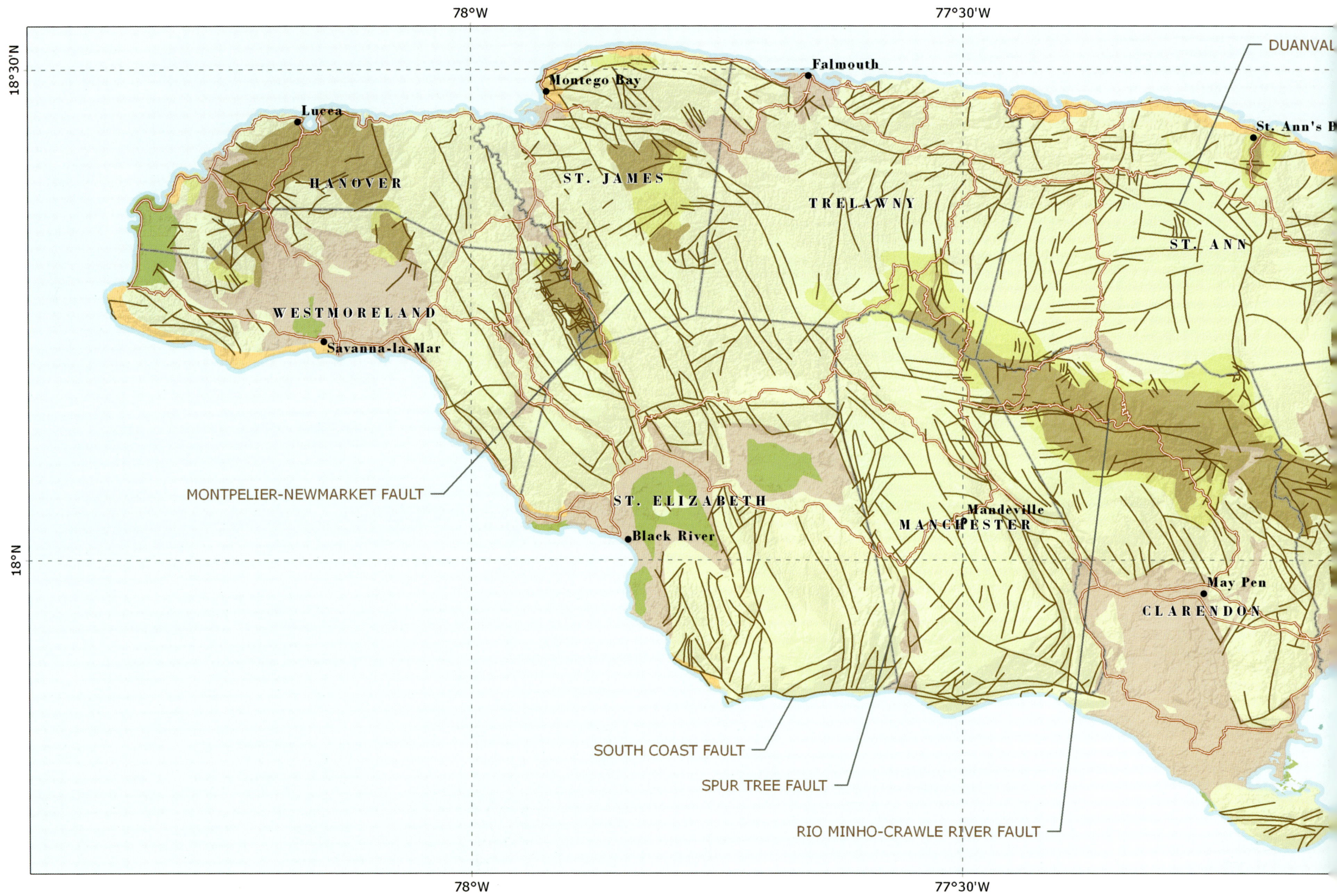

Map 1.2.1 Geological formations across Jamaica

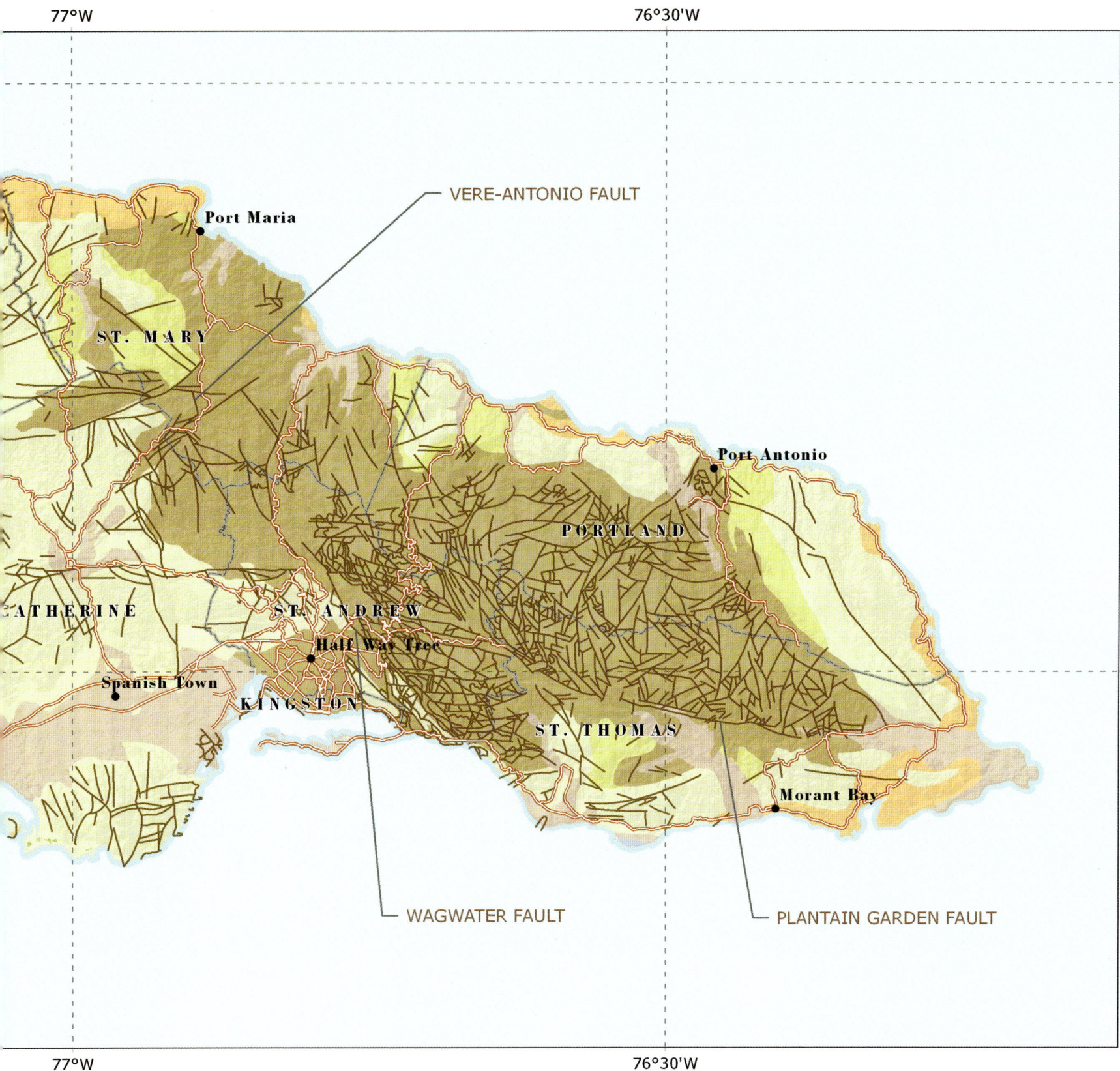

77°W · 76°30'W

18°30'N · 18°N

VERE-ANTONIO FAULT

Port Maria

ST. MARY

Port Antonio

PORTLAND

CATHERINE

ST. ANDREW

Half Way Tree

Spanish Town

KINGSTON

ST. THOMAS

Morant Bay

WAGWATER FAULT

PLANTAIN GARDEN FAULT

Geology

- Alluvium
- Coastal limestone
- White limestone
- Yellow limestone
- Volcaniclastic
- Swamp land
- Salt pond
- Coral reef
- Fault
- Parish capital
- Parish boundary
- Main road

1:450,000

0 5 10 20 kilometres

0 5 10 20 miles

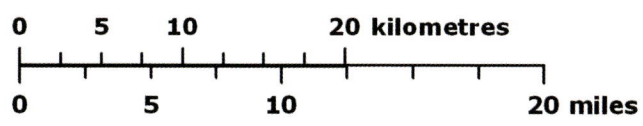

Map Datum:
Latitude-Longitude geodetic grid
World Geodetic System 1984 datum
Prime Meridian: Greenwich
Angular unit: Degree

1.3 PROTECTED AREAS

Jamaica has several protected areas distributed across the island. Invariably, these protected areas have biological significance, but all are entwined with the physical environment. The interrelationship between the natural and physical environments has a significant role in the management of natural hazards. Deforestation from forest fires, cultivation and settlement along hill slopes all lead to increased surface run-off and accelerated soil erosion, which result in increased sediment flows in river channels, which result in both water and debris flooding downstream.

Destruction of marine ecosystems, such as coral reefs and mangroves, removes natural shoreline protection against storm surges. And with the interrelationship between the natural and physical environments, landslides and over-sedimentation can cause plumes in the sea which may damage marine environments.

Table 1.3.1 Protected Areas in Jamaica (Declared and Proposed)

Protected Area	Status	Location
Black River Morass	Proposed	St Elizabeth
Cockpit Country	Proposed	Trelawny/St James
Dolphin Head Reserve	Proposed	Hanover
Port Antonio Marine Park	Proposed	Portland
Blue and John Crow Mountains National Park	Declared	Portland/St Thomas/St Andrew
Montego Bay Marine Park	Declared	St James
Negril Environmental Protected Area and Marine Park	Declared	Westmoreland
Ocho Rios Marine Park	Declared	St Ann
Palisadoes/Port Royal Protected Area	Declared	Kingston
Portland Bight Protected Area	Declared	Clarendon/St Catherine

Source: NEPA 2005.

Plate 1.21 Rabbit Hand Pass to Hermann Trapps, Carmarthen East

78°30'W   78°W   77°30'W

18°30'N

Montego Bay Marine Park

Dolphin Head Reserve

IRONSHORE

Falmouth

Montego Bay

MOUNT SALEM

DISCOVERY BAY

Lucea

ADELPHI

DUNCANS

RUNAWAY BAY

St. Ann

HANOVER

KEMPSHOT

WAKEFIELD

CLARKS TOWN

BROWNS TOWN

BAMBOO

ST. JAMES

TRELAWNY

STEWART TOWN

ST. ANN

CLAREMONT

KENDAL

MAROON TOWN

WESTMORELAND

CAMBRIDGE

ULSTER SPRING

ALEXANDRIA

MON

CATADUPA

Cockpit Country

ALBERT TOWN

YORK CASTLE

NEGRIL

TROY

WAIT-A-BIT

Savanna-la-Mar

DARLISTON

Negril Environmental Protected Area and Marine Park

GINGER HILL

ACCOMPONG

BALACLAVA

CHRISTIANA

KELLITS

NEWMARKET

MAGGOTTY

LLUIDAS VALE

ST. ELIZABETH

BUSHY PARK

MIDDLE QUARTERS

KENDAL

THOMPSON TOWN

CHAPELTON

LACOVIA

SANTA CRUZ

WILLIAMSFIELD

Black River

PORUS

MOCHO

18°N

Mandeville

CLARENDON

PAROTTEE

MALVERN

SPUR TREE

NAIN

NEWPORT

May Pen

HALSE HALL

Black River Morass

MANCHESTER

JUNCTION

BULL SAVANNA

SANDY BAY

LIONEL TOWN

ALLIGATOR POND

PORTLAND COTTAGE

17°30'N

Portland Bi Protected A

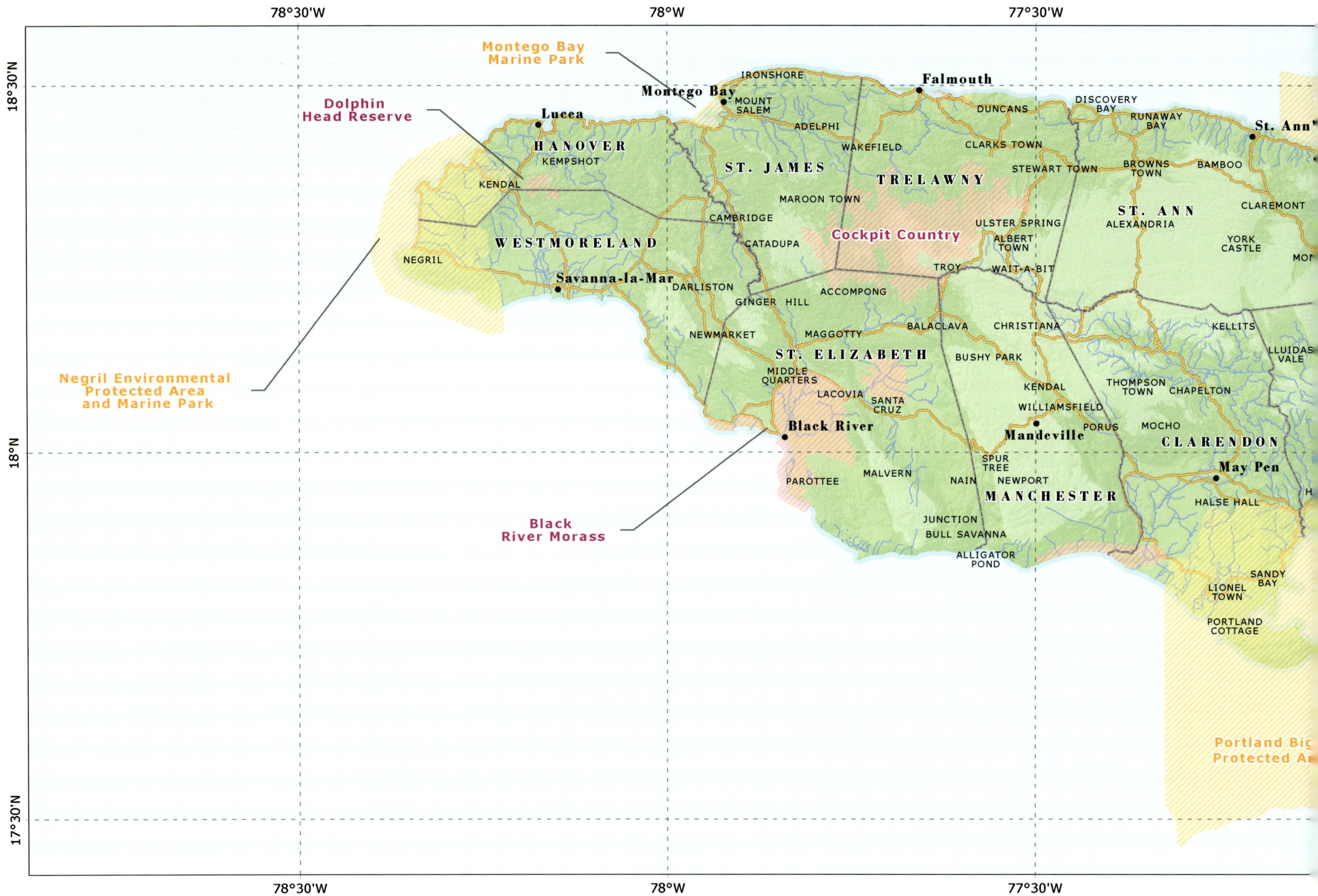

Map 1.3.1 Declared and proposed protected areas of Jamaica

– 16 –

Protected area

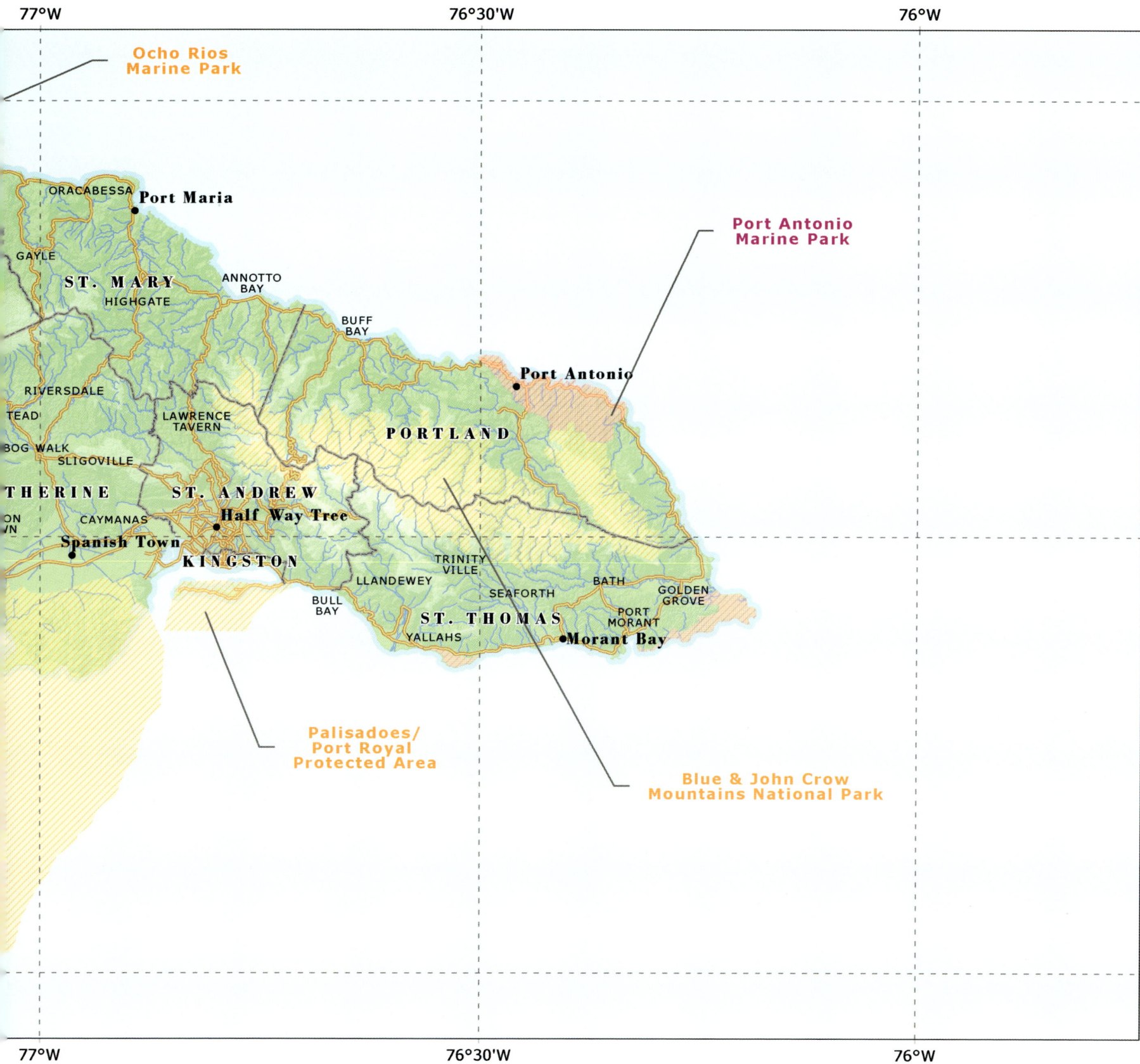

- Declared
- Proposed
- Parish capital
- Parish boundary
- Main road
- Major river

Elevation (metres)

- 0—500
- 501—1,000
- 1,001—2,258

Ocho Rios
Marine Park

Port Antonio
Marine Park

Palisadoes/
Port Royal
Protected Area

Blue & John Crow
Mountains National Park

ORACABESSA
Port Maria
GAYLE
ST. MARY
HIGHGATE
ANNOTTO
BAY
BUFF
BAY
Port Antonio
RIVERSDALE
LAWRENCE
TAVERN
PORTLAND
TEAD
BOG WALK
SLIGOVILLE
THERINE
ST. ANDREW
CAYMANAS
Half Way Tree
Spanish Town
KINGSTON
TRINITY
VILLE
LLANDEWEY
SEAFORTH
BATH
GOLDEN
GROVE
BULL
BAY
PORT
MORANT
ST. THOMAS
YALLAHS
Morant Bay

77°W
76°30'W
76°W
18°30'N
18°N
17°30'N

1:602,734

| 0 | 5 | 10 | | 20 kilometres |

| 0 | 5 | 10 | | 20 miles |

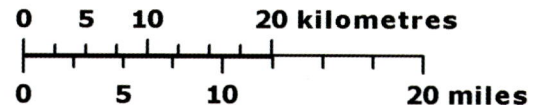

Map Datum:
Latitude-Longitude geodetic grid
World Geodetic System 1984 datum
Prime Meridian: Greenwich
Angular unit: Degree

Plate 1.3.2 Portland Bight Protected Area (P. Lyew-Ayee Jr, 2006)

1.4 RAINFALL

Jamaica has a distinct hydrometeorological profile. In the eastern sections, the presence of the Blue Mountains promotes orographic rainfall on the northern slopes in Portland, as well as a rain shadow on the mountains' southern flanks. Much of the southern parishes also have relatively lower rainfall compared to other parts of Jamaica.

Jamaica's hydrogeological character is also closely connected. In the volcaniclastic Blue Mountains, rainfall leads to surface run-off, which is responsible for the sculpting of the landscape and creation of many river systems in the east, such as the Rio Grande, Yallahs River, Morant River and the Plantain Garden River. In periods of low rainfall, the flow in these rivers is drastically reduced, adversely affecting water supply in these areas. Elsewhere, in the limestone regions, groundwater collection is much more pronounced, and underground water storage ensures sustained flows and supplies in these areas, even during periods of low rainfall.

Plate 1.4.1 Flood damage, Hope River at Kintyre, St Andrew, Tropical Storm Nicole, 2010 (L. Buchanan, 2010)

Table 1.4.1 Jamaica's Hydrometeorological Profile

Highest 30-year mean rainfall for a parish	Portland
Lowest 30-year mean rainfall for a parish	Clarendon
Highest 30-year mean rainfall for a town	Bowden Pen, Portland
Lowest 30-year mean rainfall for a town	Bull Bay, St Andrew
Rainiest month in Jamaica	October
Driest month in Jamaica	March

Source: Meteorological Service of Jamaica.

Plate 1.4.2 Flood damage to houses on the western bank of the Hope River at Kintyre, St Andrew, Tropical Storm Nicole, 2010 (L. Buchanan, 2010)

Plate 1.4.3 Flood damage to fording, Hope River at Kintyre, St Andrew, Tropical Storm Nicole, 2010 (L. Buchanan, 2010)

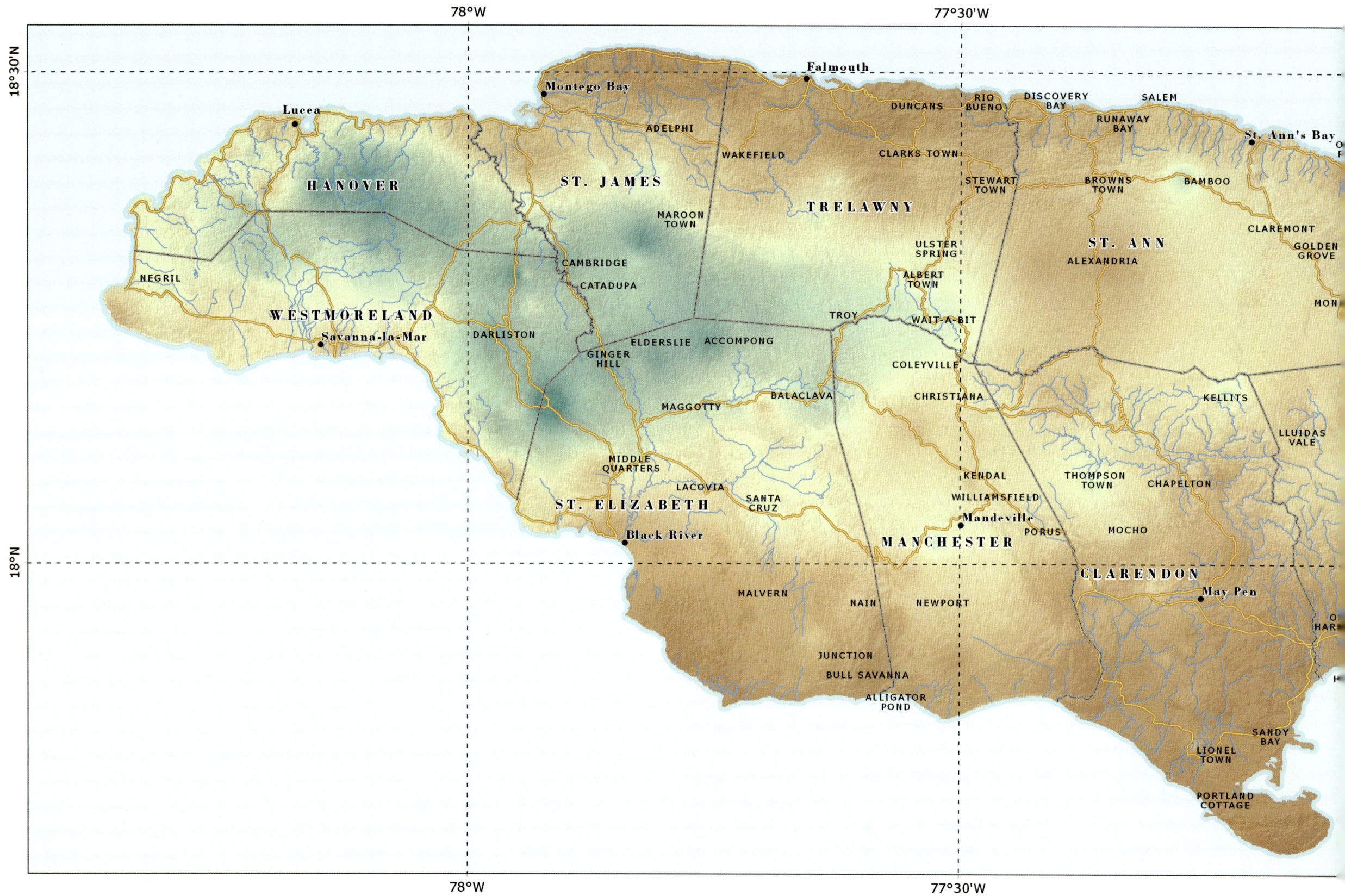

Map 1.4.1 Thirty-year mean rainfall patterns across Jamaica

18°30'N

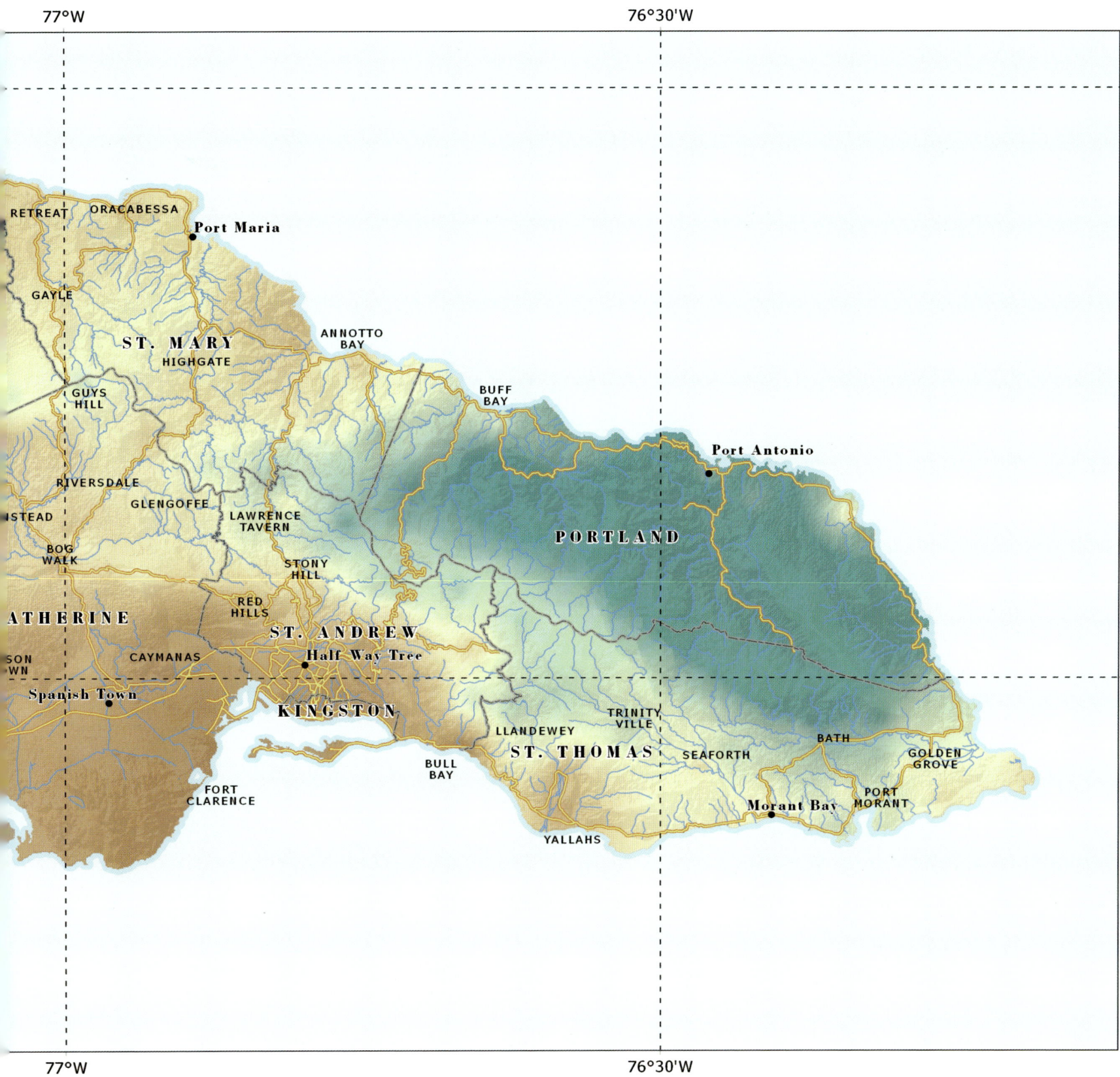

30-year mean rainfall (mm)

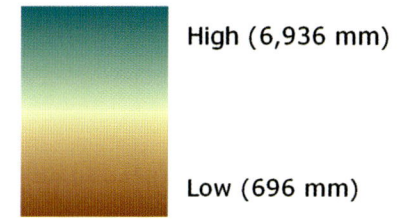

High (6,936 mm)

Low (696 mm)

- Parish capital

Parish boundary

Main road

Major river

RETREAT   ORACABESSA
Port Maria

GAYLE

ST. MARY
HIGHGATE

ANNOTTO
BAY

GUYS
HILL

BUFF
BAY

RIVERSDALE

GLENGOFFE

LAWRENCE
TAVERN

Port Antonio

STEAD

BOG
WALK

STONY
HILL

PORTLAND

ATHERINE

RED
HILLS

ST. ANDREW
Half Way Tree

SON
WN

CAYMANAS

18°N

Spanish Town

KINGSTON

LLANDEWEY

TRINITY
VILLE

BATH

BULL
BAY

ST. THOMAS

SEAFORTH

GOLDEN
GROVE

FORT
CLARENCE

Morant Bay

PORT
MORANT

YALLAHS

18°N

1:450,000

0	5	10	20 kilometres

0	5	10	20 miles

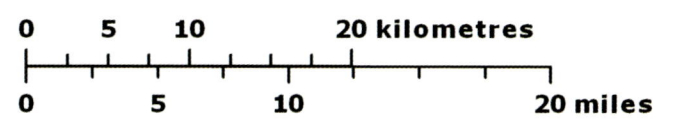

Map Datum:
Latitude-Longitude geodetic grid
World Geodetic System 1984 datum
Prime Meridian: Greenwich
Angular unit: Degree

1.5 HYDROLOGY

Jamaica's hydrology is closely tied to its geology. Volcaniclastic areas form aquicludes, where the impermeable rocks promote surface run-off, directing flows toward rivers and streams. Eastern Jamaica has a much greater concentration of rivers than the rest of the island, its sculpted landscape the result of these hydrological processes. There are other smaller aquicludes in central and western Jamaica; all are associated with the headwaters of major river systems.

The limestone regions promote groundwater storage, usually at great depth; these issue forth as springs when the water table is close to the surface, especially closer to the coast. Many of the shorter rivers on the north coast of Jamaica begin as springs coming from the limestone aquifers, including the Dunn's River in St Ann. Some alluvial plains serve as alluvium aquifers, where the sediment promotes water storage.

Both ground and surface water are extremely sensitive to pollution and are entirely replenished from rainwater; there is no outside source for freshwater on the island, and there is no desalination plant serving Jamaica.

Table 1.5.1 Average River Flows

River	Flow (cubic metres per second)
Rio Grande	25.0
Black River	19.7
Great River	12.2
Rio Cobre	10.7
Martha Brae	8.4
Wagwater	6.5

Source: Water Resources Authority, 2011

Plate 1.5.1 Rio Minho (Water Resources Authority, 2002)

Plate 1.5.2 Upper Hope River (R. Ahmad, 2003)

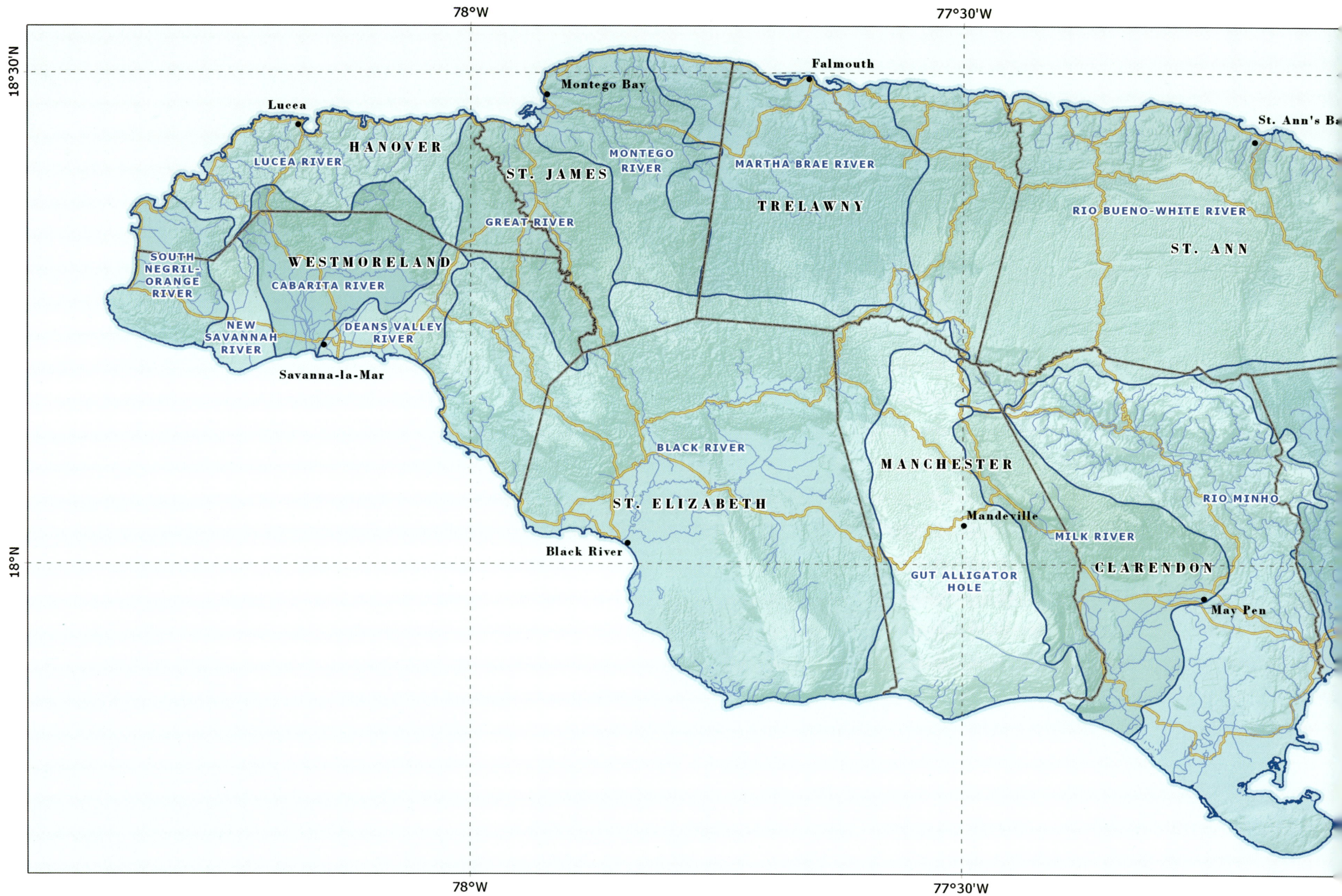

Map 1.5.1 Hydrological basins across Jamaica

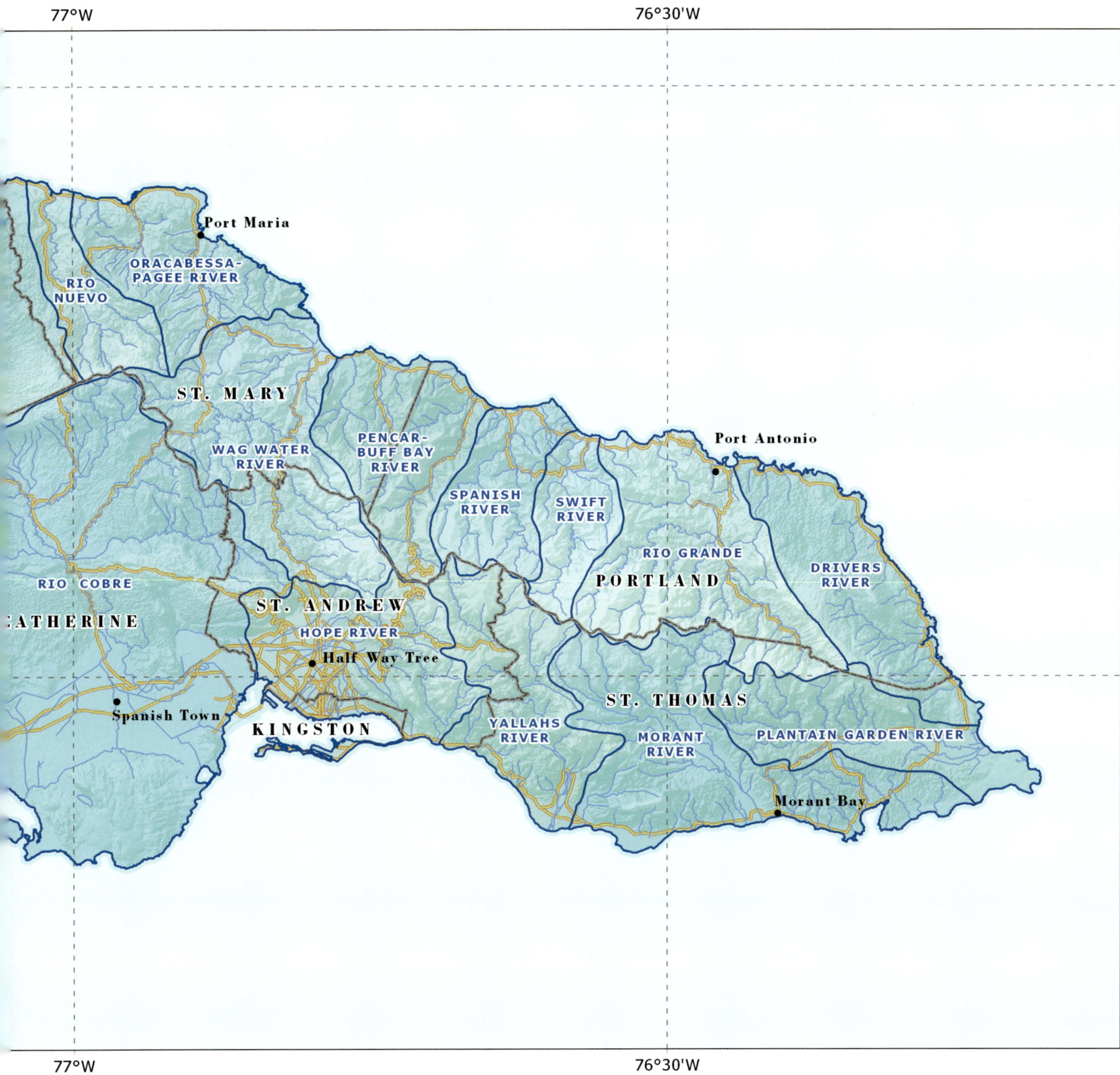

18°30'N

Watershed boundary

Parish capital

Parish boundary

Main road

Major river

Port Maria

ORACABESSA-
PAGEE RIVER

RIO
NUEVO

ST. MARY

WAG WATER
RIVER

PENCAR-
BUFF BAY
RIVER

SPANISH
RIVER

SWIFT
RIVER

Port Antonio

RIO GRANDE

DRIVERS
RIVER

PORTLAND

RIO COBRE

ST. ANDREW

HOPE RIVER

Half Way Tree

18°N

:ATHERINE

ST. THOMAS

Spanish Town

YALLAHS
RIVER

MORANT
RIVER

PLANTAIN GARDEN RIVER

KINGSTON

Morant Bay

1:450,000

| 0 | 5 | 10 | 20 kilometres |

| 0 | 5 | 10 | 20 miles |

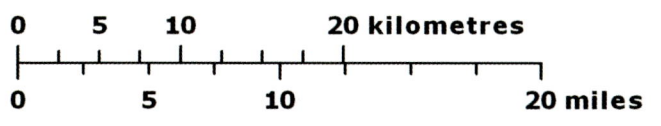

Map Datum:
Latitude-Longitude geodetic grid
World Geodetic System 1984 datum
Prime Meridian: Greenwich
Angular unit: Degree

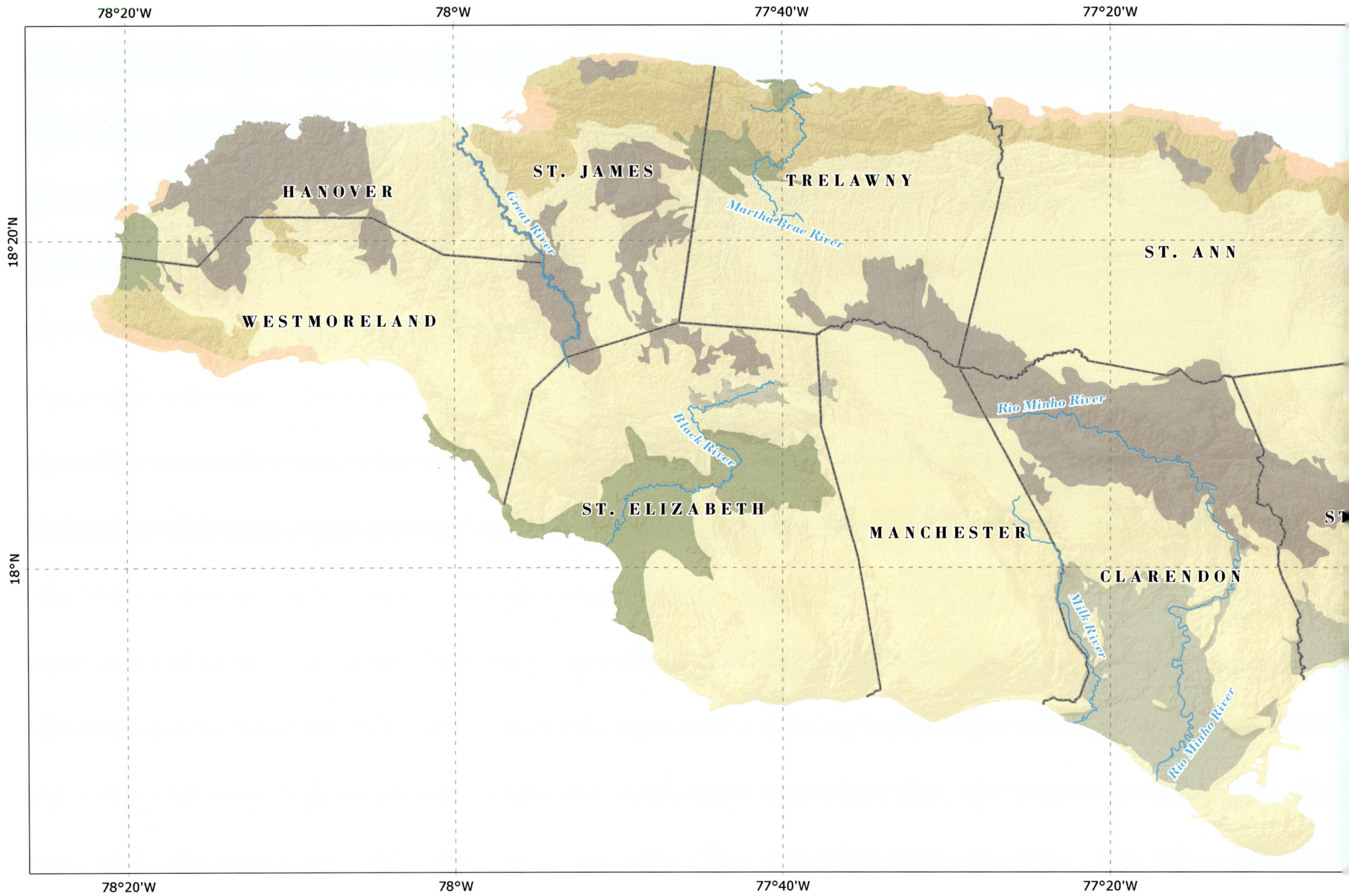

Map 1.5.2 Hydrostratigraphy of Jamaica

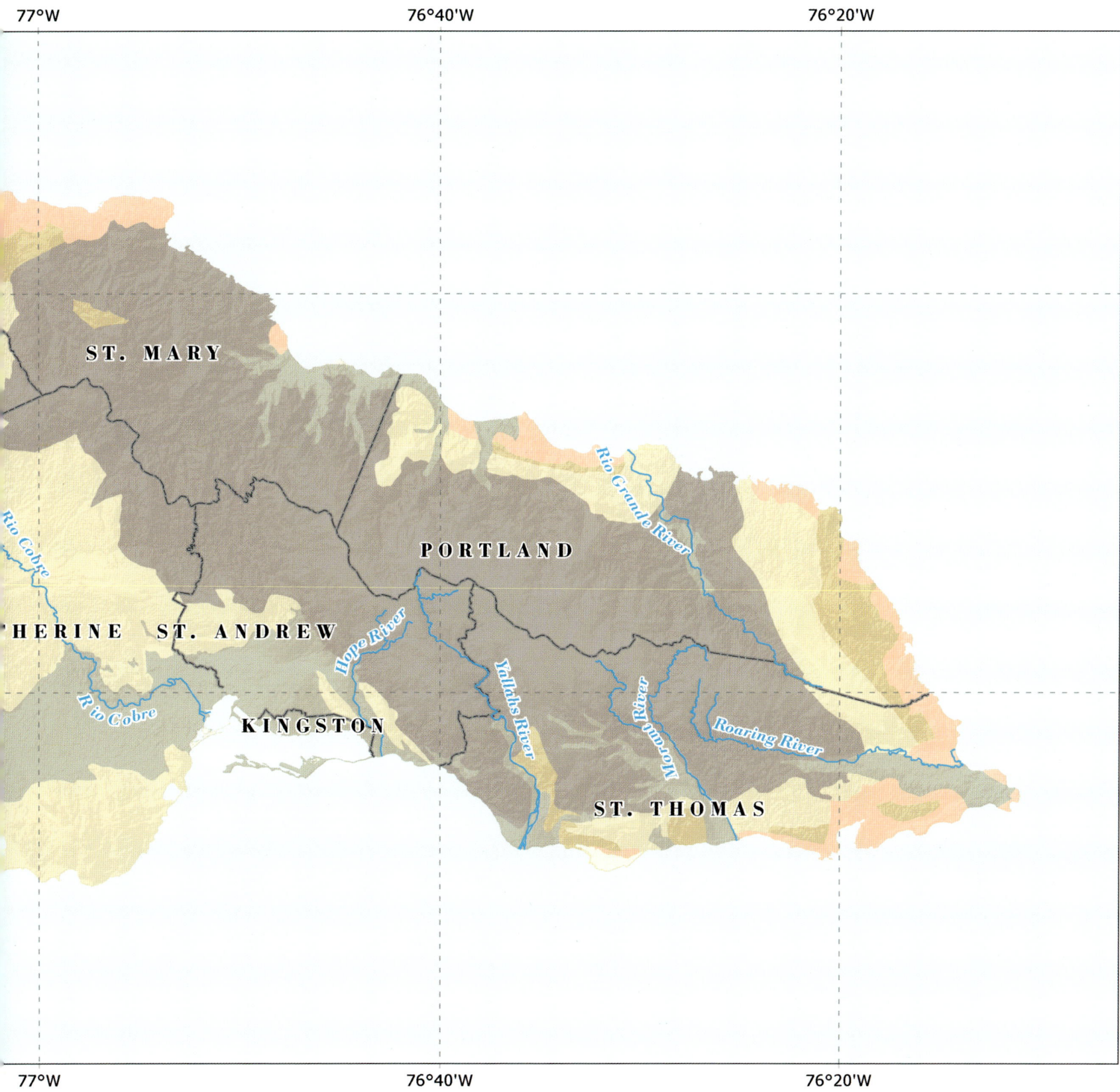

Hydrostatigraphy

- Alluvium aquiclude
- Alluvium aquifer
- Basal aquiclude
- Coastal aquiclude
- Limestone aquiclude
- Limestone aquifer
- Water course
- Parish boundary
- Major river

ST. MARY

PORTLAND

HERINE ST. ANDREW

Hope River

Rio Grande River

Rio Cobre

Rio Cobre

KINGSTON

Yallahs River

Morant River

Roaring River

ST. THOMAS

1:450,000

| 0 | 5 | 10 | 20 kilometres |

| 0 | 5 | 10 | 20 miles |

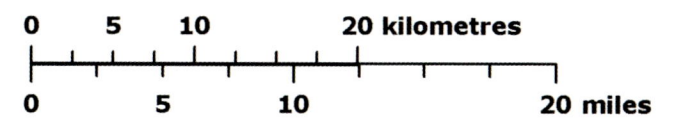

Map Datum:
Latitude-Longitude geodetic grid
World Geodetic System 1984 datum
Prime Meridian: Greenwich
Angular unit: Degree

77°W   76°40'W   76°20'W

18°20'N

18°N

Facts and Figures

NATURAL HAZARDS

1. Jamaica is vulnerable to four main natural hazards: tropical storms/hurricanes, earthquakes, floods and landslides.

2. Flooding is the most common hazard, followed by landslides.

3. There are two broad types of flooding: coastal and inland.

4. Flooding and landslides may occur as secondary hazards from primary events such as hurricanes and earthquakes, as well as from man-made triggers.

5. Multiple hazards pose significant threats to communities. The five multihazard-prone communities (by composite percentage of land area vulnerable to each natural hazard type) are:

    • Orange Bay, St Thomas

    • Southside, Kingston

    • Independence City, St Catherine

    • Westchester, St Catherine

    • Westmeade, St Catherine

2.1 RISK PROFILE

Jamaica is exposed to four major hazards: hurricanes, earthquakes, floods and landslides. While hurricanes and earthquakes are associated with the most spectacular natural disasters (June 1692 and January 1907 earthquakes; September 1988 Hurricane Gilbert), floods and landslides are the most frequent and abundant hazards. Only hurricanes have a defined season; earthquakes, floods and landslides may occur at any time.

Floods and landslides can be secondary disasters, occurring as a result of hurricane or earthquake impacts; hurricanes are primarily windstorm events, while earthquakes are ground-shaking events. These may also occur as a result of man-made activities (broken pipelines, poor construction, hill-blasting and the like). However, once these activities trigger flooding or landslides, natural processes (gravity, topography and so on) influence the effects and extent of the impact.

This atlas looks at two broad types of flooding: coastal and inland. Low-lying coastal terrain, especially along Jamaica's south coast, is prone to flooding, either sudden and catastrophic (storm surges or tsunamis) or slow and gradual (coastal erosion, sea level rise and the like). Inland flooding may occur in gently sloping inland regions as a result of riverine flooding or surface run-off due to the inability of surface water to be channelled to proper drainage channels or redirected underground, due to blocked drains, saturated soils or paved surfaces. Typically, river overflow or surface run-off occurs as a result of intense and/or prolonged rainfall events.

Similar events may also trigger landslides, where water-laden soil, coupled with gravity and geological structure, moves. Landslides may also occur as a result of ground shaking brought about by earthquakes.

Hazard impacts have considerable effects on Jamaica's environment, economy, infrastructure, society, individuals and government.

2.2 HURRICANES

Jamaica is vulnerable to tropical storms and hurricanes during the Atlantic hurricane season, which runs from June through November each year. During this period, warm sea surface temperatures promote the development of lower atmospheric circulations which form into tropical storms and hurricanes. A total of sixty-one recorded systems have come within 100 kilometres of Jamaica since 1851.

Tropical storms have sustained wind speeds of up to 120 kilometres per hour (74 miles per hour), while the strongest hurricanes have wind speeds in excess of 185 kilometres per hour (155 miles per hour). Secondary effects of tropical storms and hurricanes include floods and landslides.

At a local scale, topography plays an important role in focusing wind effects. Inland plains provide little topographic resistance to winds, although surface friction may mitigate against top speeds. Upland areas, both limestone and non-limestone, may be more exposed, depending on the storm's approach, although the leeward slopes would be less exposed. Depending on their alignments, river valleys may channel and accelerate wind speeds.

Most of Jamaica's buildings are situated on the gently sloping plains, particularly along the Liguanea Plain in Kingston, and the Caymanas Plain in St Catherine. Buildings situated in enclosed valleys may be more sheltered from high winds during a hurricane event, while buildings on exposed slopes are at higher risk. Construction type now becomes an important factor in determining each building's vulnerability to damage during an event.

During a single event, therefore, all buildings in Jamaica will not necessarily experience uniform storm conditions, and even in places experiencing similar conditions, damage may be uneven due to different construction types.

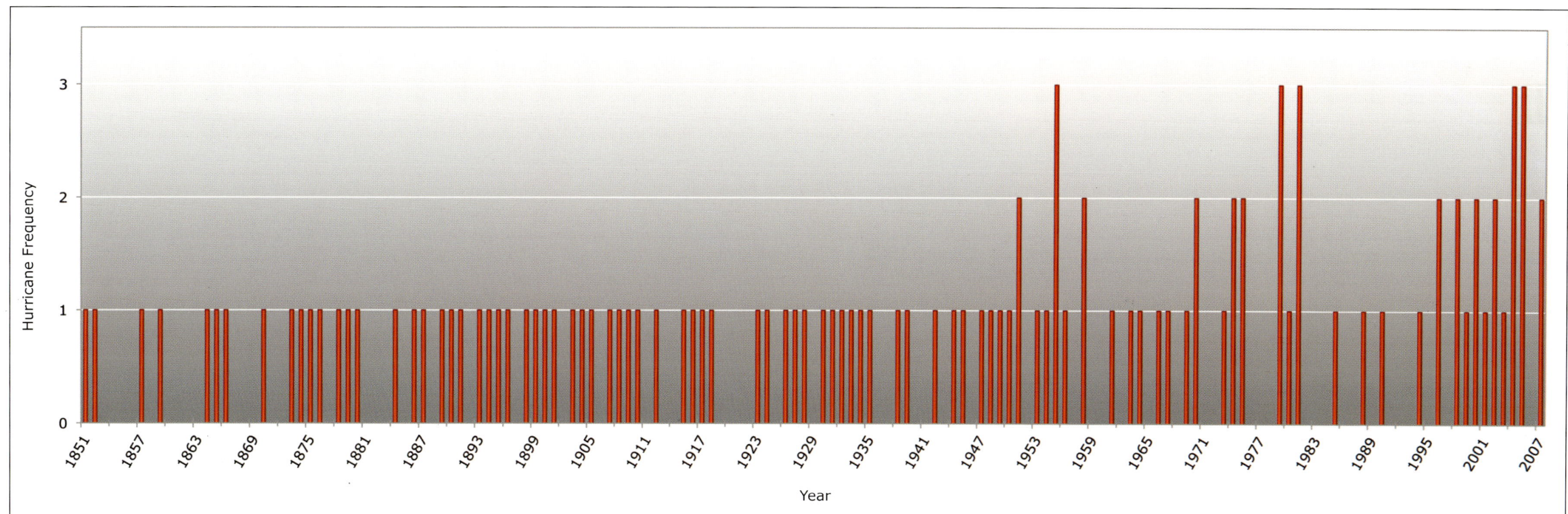

Graph 2.2.1 Hurricane frequency graph

Table 2.2.1 Hurricane Facts and Figures

Total number of the strongest storm systems within 100 kilometres of Jamaica since 1851	61
Total number of storm systems making landfall	23
Highest number of storms in a single year	3 (2004)
Highest number of storms in a single month for the period	20 (September) 16 (August) 12 (October) 6 (November) 3 (July) 2 (June) 1 (May) 1 (December)
Strongest storm	Ivan (2004)

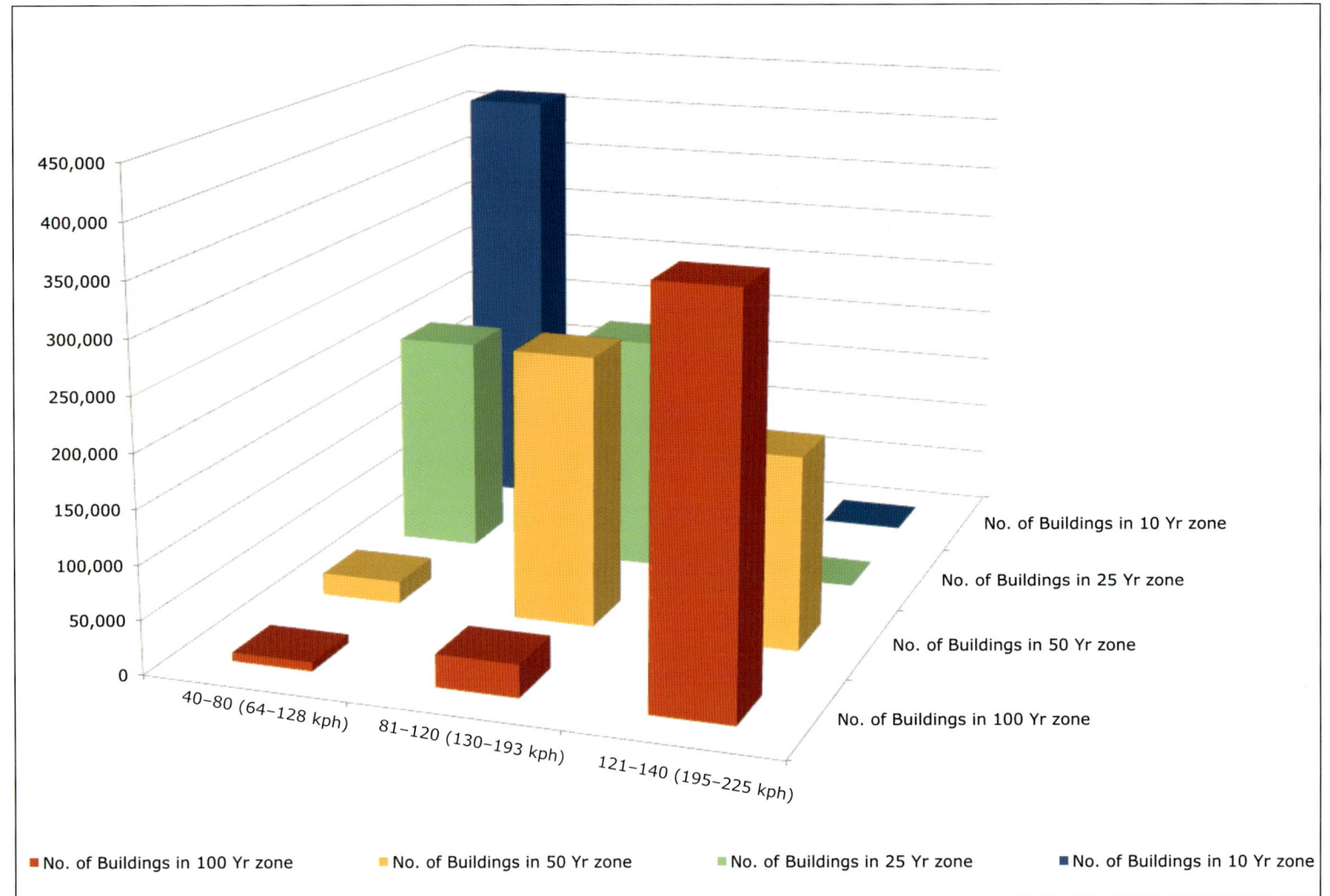

Graph 2.2.2 Built environment and hurricane return

Table 2.2.2 Number of Buildings Impacted by Hurricanes over a 100-Year Period in Jamaica

Wind speed (mph)	No. of buildings in 100-year-zone	No. of buildings in 50-year-zone	No. of buildings in 25-year-zone	No. of buildings in 10-year-zone
40–80 (64–128 kph)	8,004	20,673	203,517	412,614
81–120 (130–193 kph)	30,085	251,973	220,117	47,123
121–140 (195–225 kph)	371,281	178,999	0	0

Plate 2.2.2 Damage resulting from Hurricane Ivan, Portland Cottage, Clarendon (Water Resources Authority, 2004)

Plate 2.2.3 Damage resulting from Hurricane Ivan, Portland Cottage, Clarendon (Water Resources Authority, 2004)

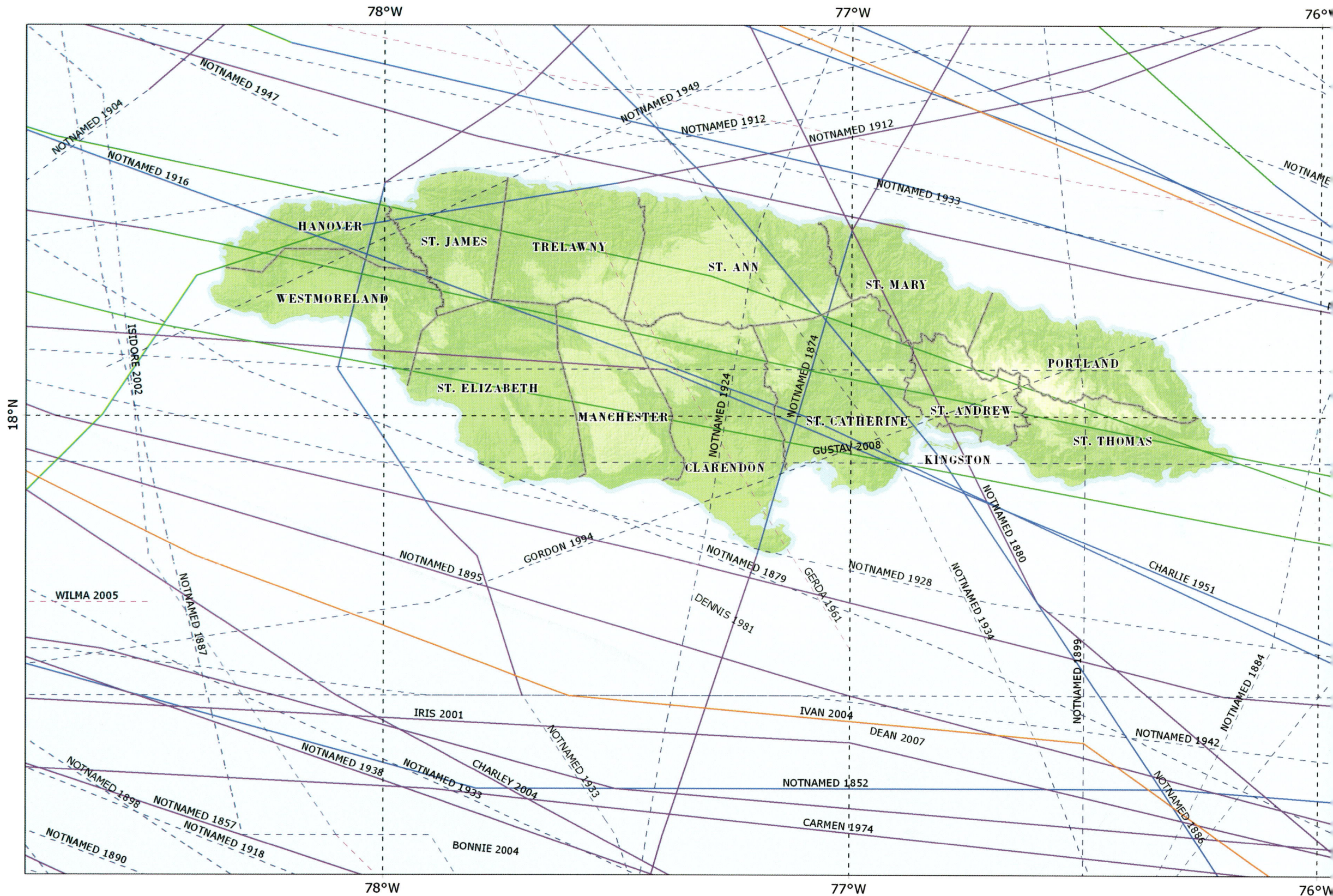

Map 2.2.1 Hurricane pathways, 1851–2006

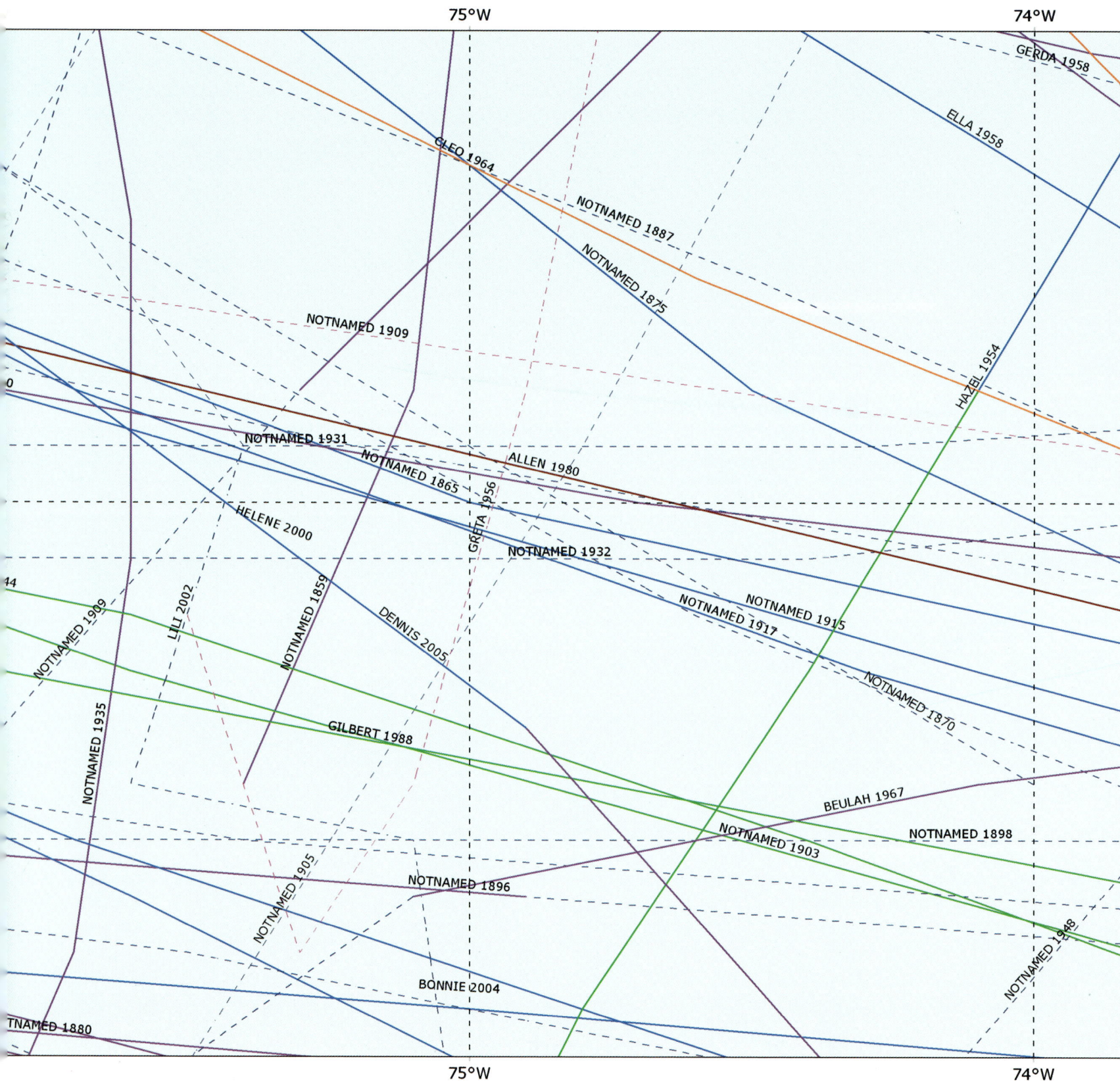

Hurricanes

- - - - Tropical low
- - - - Tropical depression
- - - - Tropical storm
———— Category 1 hurricane
———— Category 2 hurricane
———— Category 3 hurricane
———— Category 4 hurricane
———— Category 5 hurricane
———— Parish boundary

Elevation (metres)

- 0—500
- 501—1,000
- 1,001—2,258

1:950,000

| 0 | 10 | 20 | | 40 kilometres |

| 0 | 10 | 20 | | 40 miles |

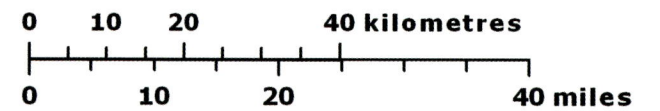

Map Datum:
Latitude-Longitude geodetic grid
World Geodetic System 1984 datum
Prime Meridian: Greenwich
Angular unit: Degree

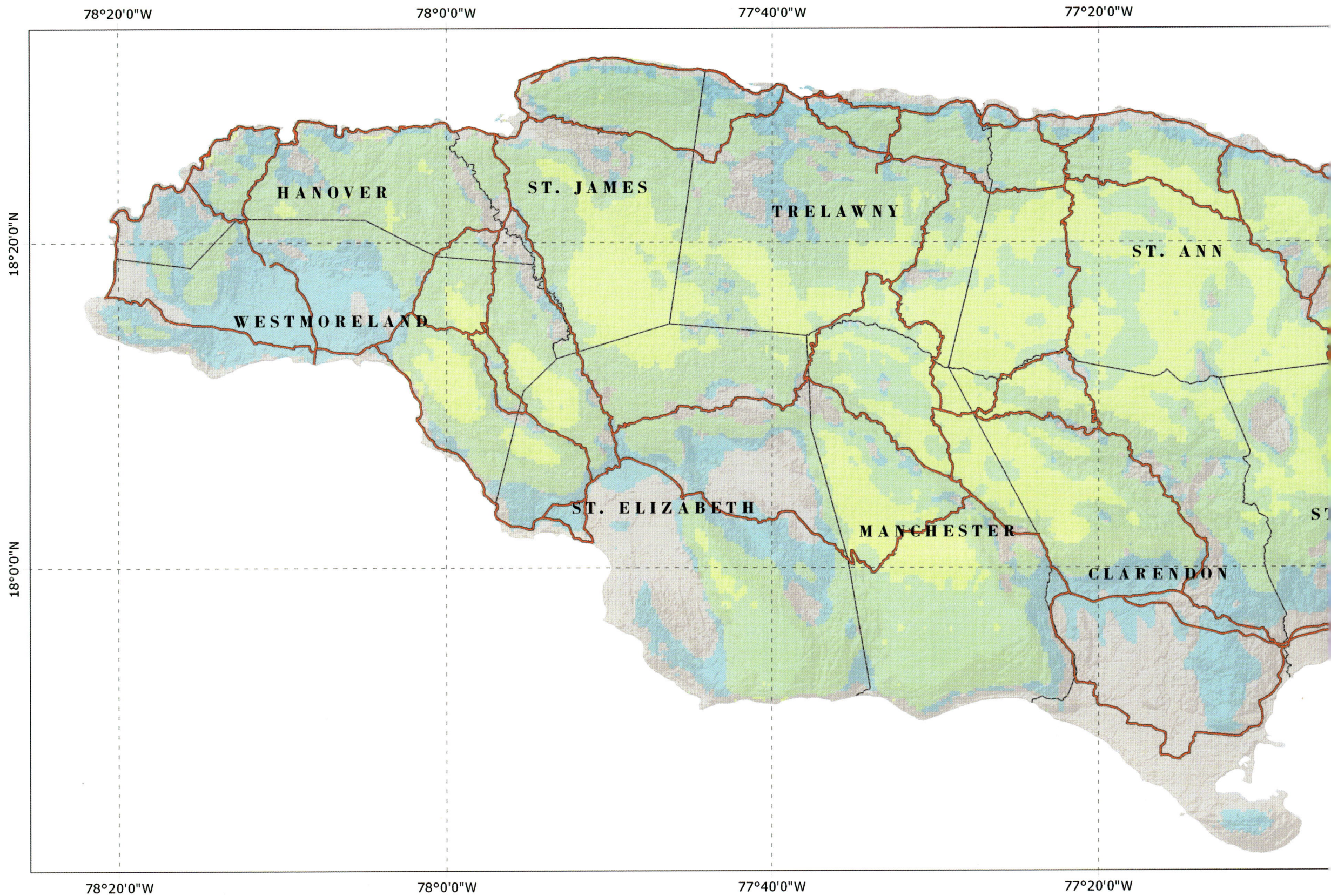

Map 2.2.2a Wind speed, 10-year return period

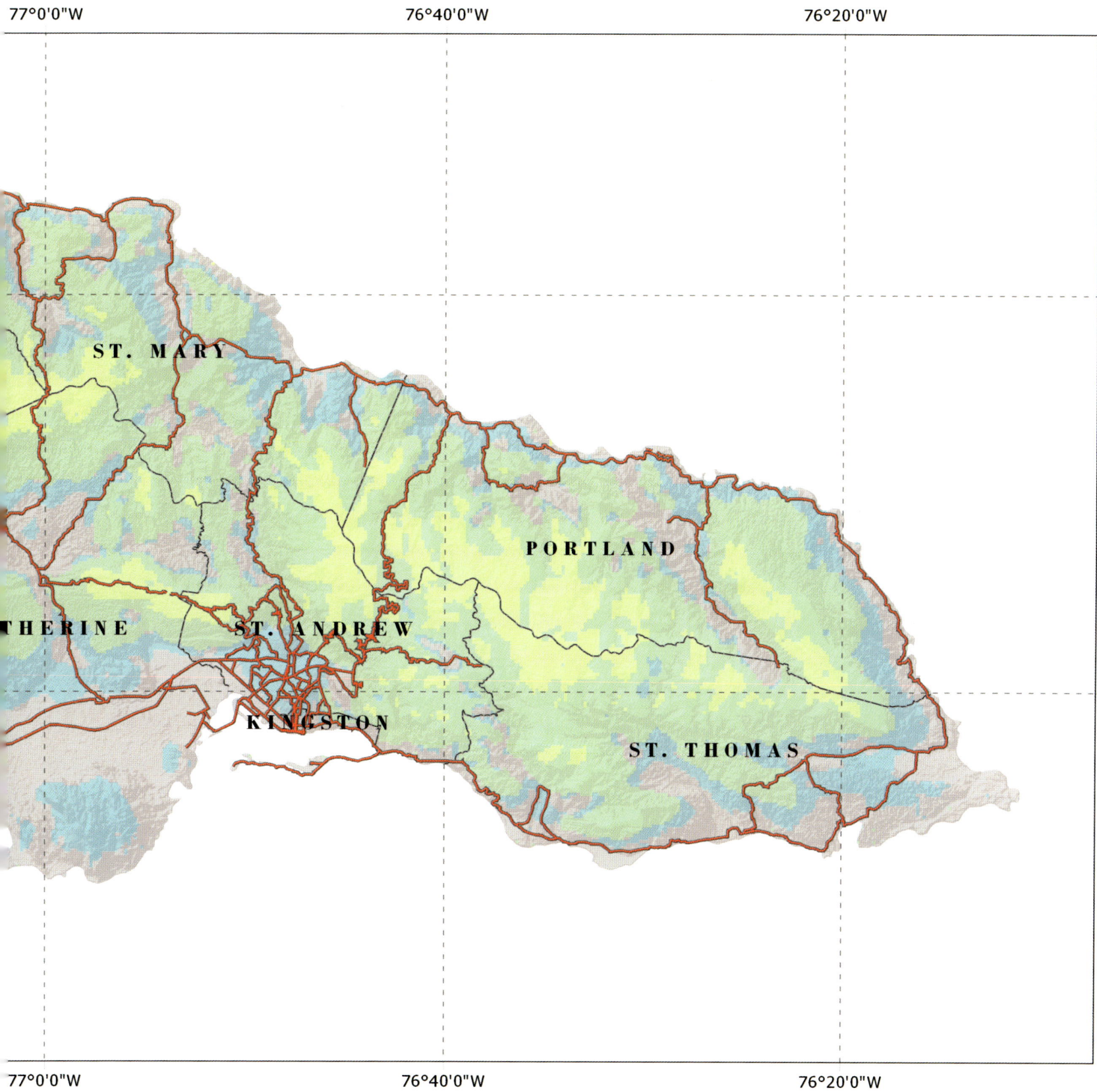

10-year return model

Wind speed (mph)

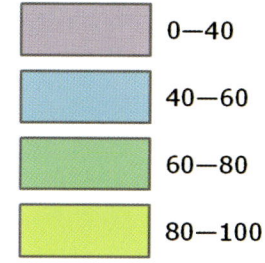

- 0—40
- 40—60
- 60—80
- 80—100

- - - - Parish boundary

—— Main road

77°0'0"W   76°40'0"W   76°20'0"W

18°20'0"N

18°0'0"N

ST. MARY

THERINE

PORTLAND

ST. ANDREW

KINGSTON

ST. THOMAS

1:450,000

0   4.5   9   18 kilometres

0   5   10   20 miles

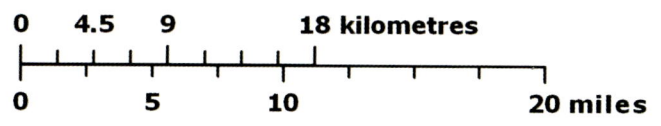

Map Datum:
Latitude-Longitude geodetic grid
World Geodetic System 1984 datum
Prime Meridian: Greenwich
Angular unit: Degree

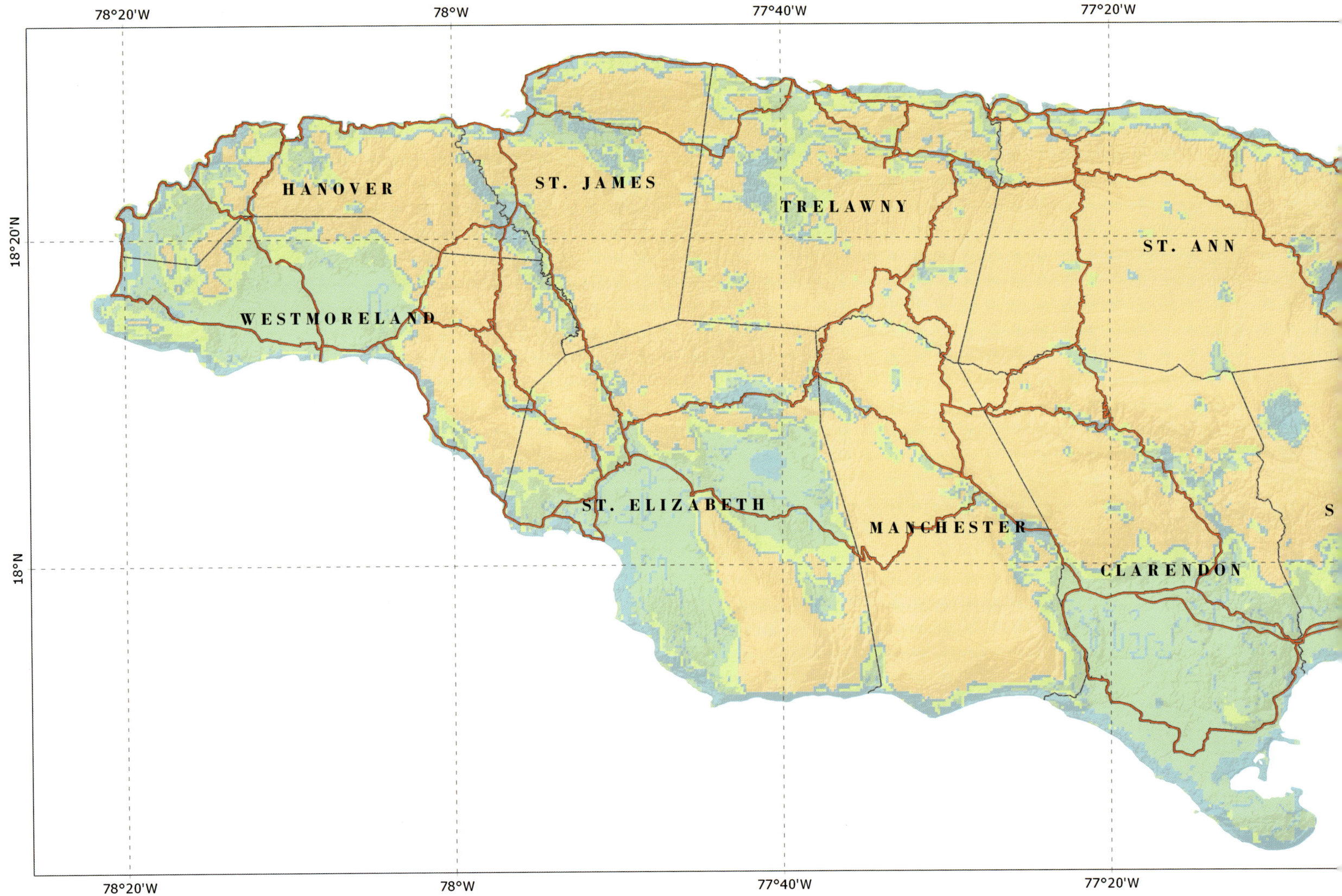

ST. JAMES

TRELAWNY

ST. ANN

WESTMORELAND

ST. ELIZABETH

MANCHESTER

CLARENDON

Map 2.2.2b Wind speed, 25-year return period

25-year return model

Wind speed (mph)

40—60

60—80

80—100

100—120

Parish boundary

Main road

18°20'N

ST. MARY

PORTLAND

THERINE

ST. ANDREW

18°N

KINGSTON

ST. THOMAS

1:450,000

| 0 | 5 | 10 | 20 kilometres |

| 0 | 5 | 10 | 20 miles |

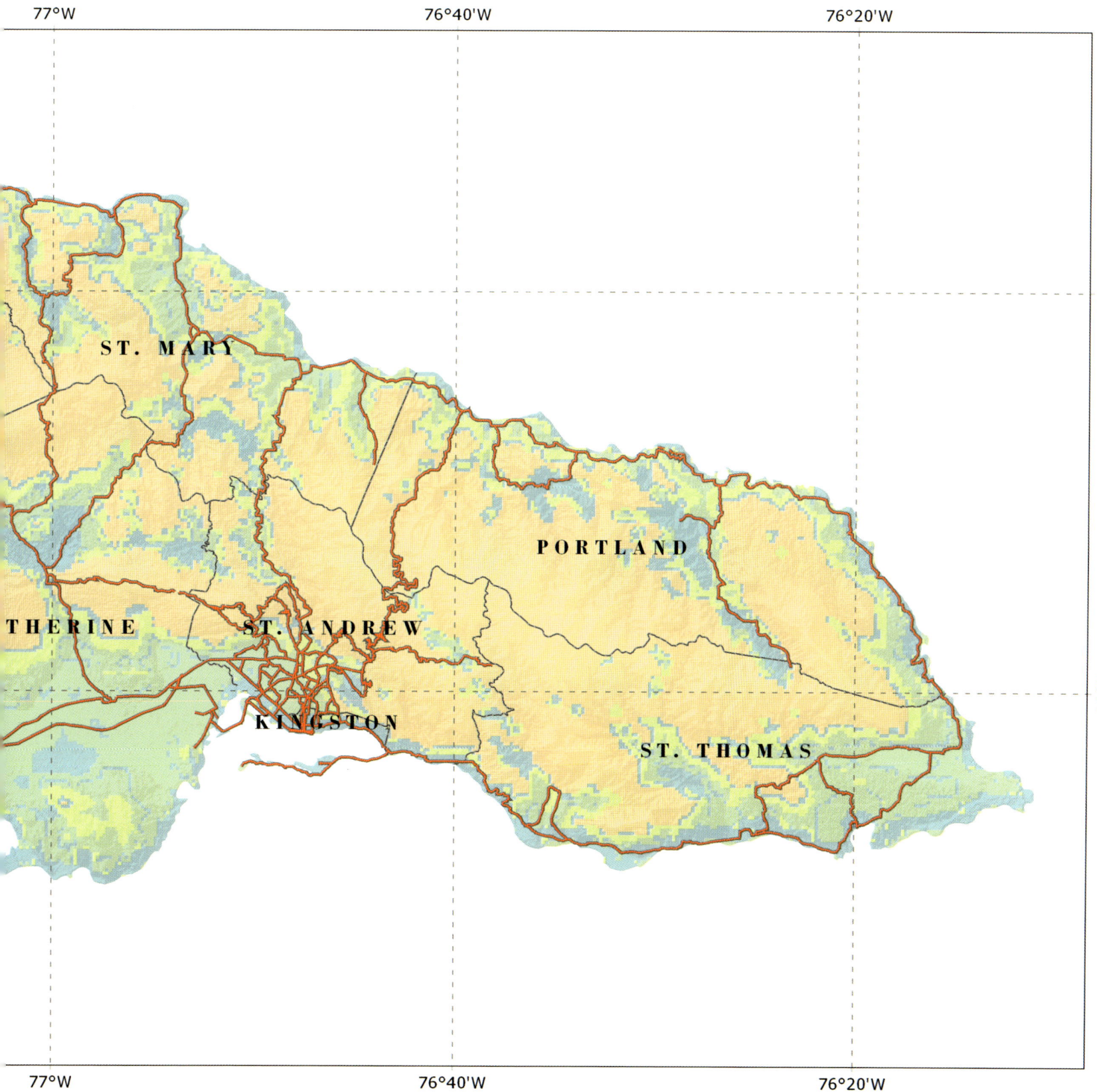

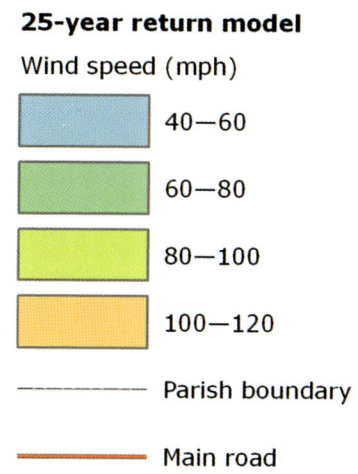

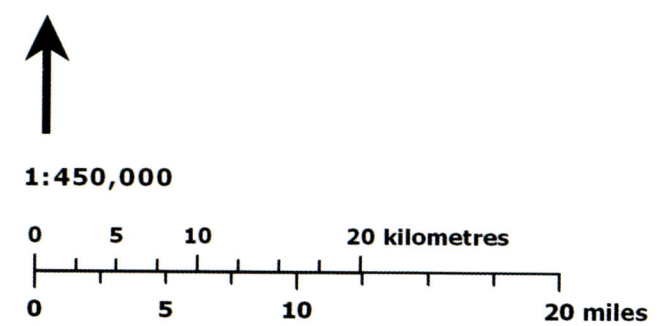

Map Datum:
Latitude-Longitude geodetic grid
World Geodetic System 1984 datum
Prime Meridian: Greenwich
Angular unit: Degree

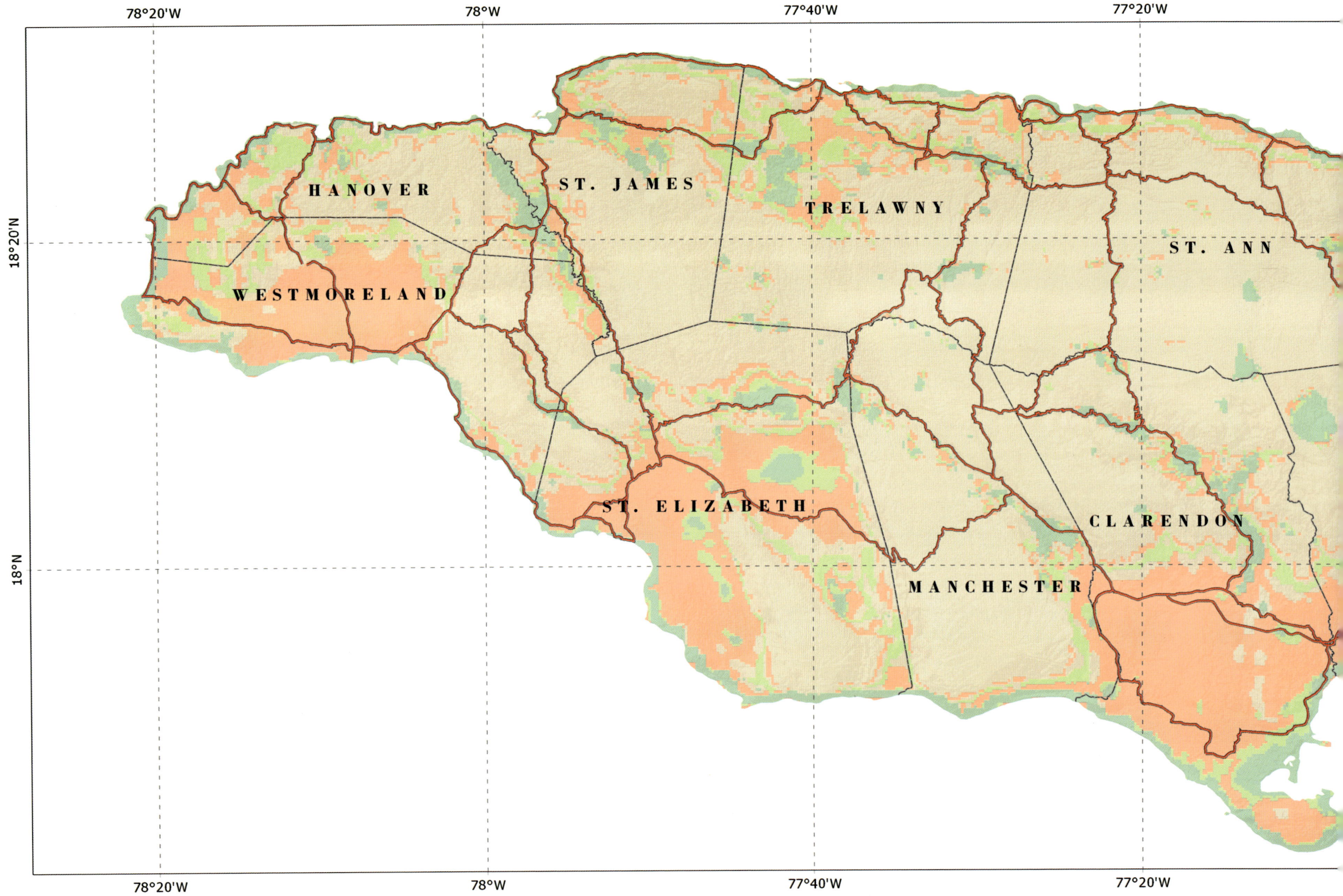

Map 2.2.2c Wind speed, 50-year return period

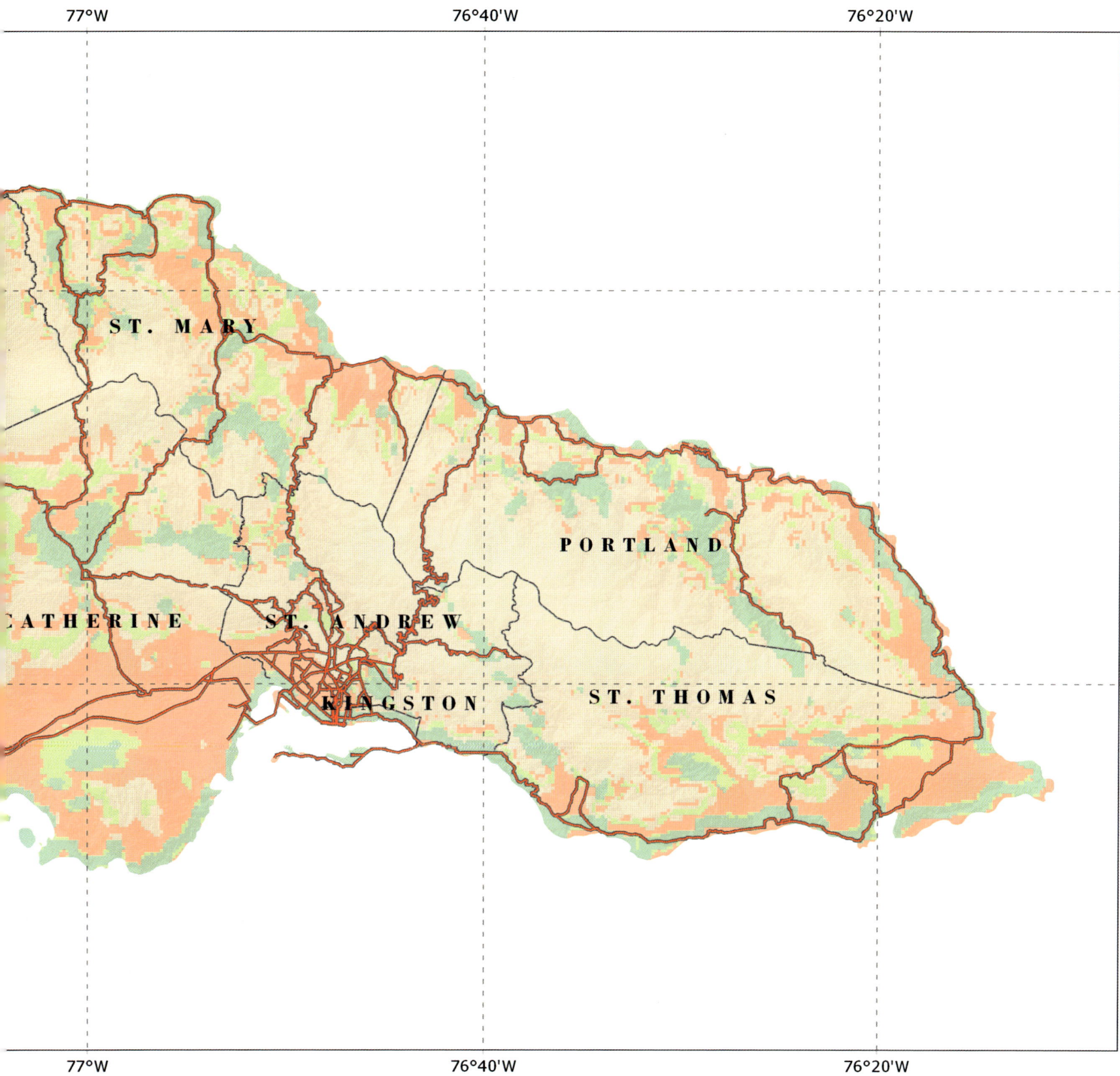

50-year return model

Wind speed (mph)

- 60—80
- 80—100
- 100—120
- 120—140

Parish boundary

Main road

18°20'N

18°N

ST. MARY

PORTLAND

CATHERINE

ST. ANDREW

KINGSTON

ST. THOMAS

1:450,000

| 0 | 5 | 10 | 20 kilometres |

| 0 | 5 | 10 | 20 miles |

Map Datum:
Latitude-Longitude geodetic grid
World Geodetic System 1984 datum
Prime Meridian: Greenwich
Angular unit: Degree

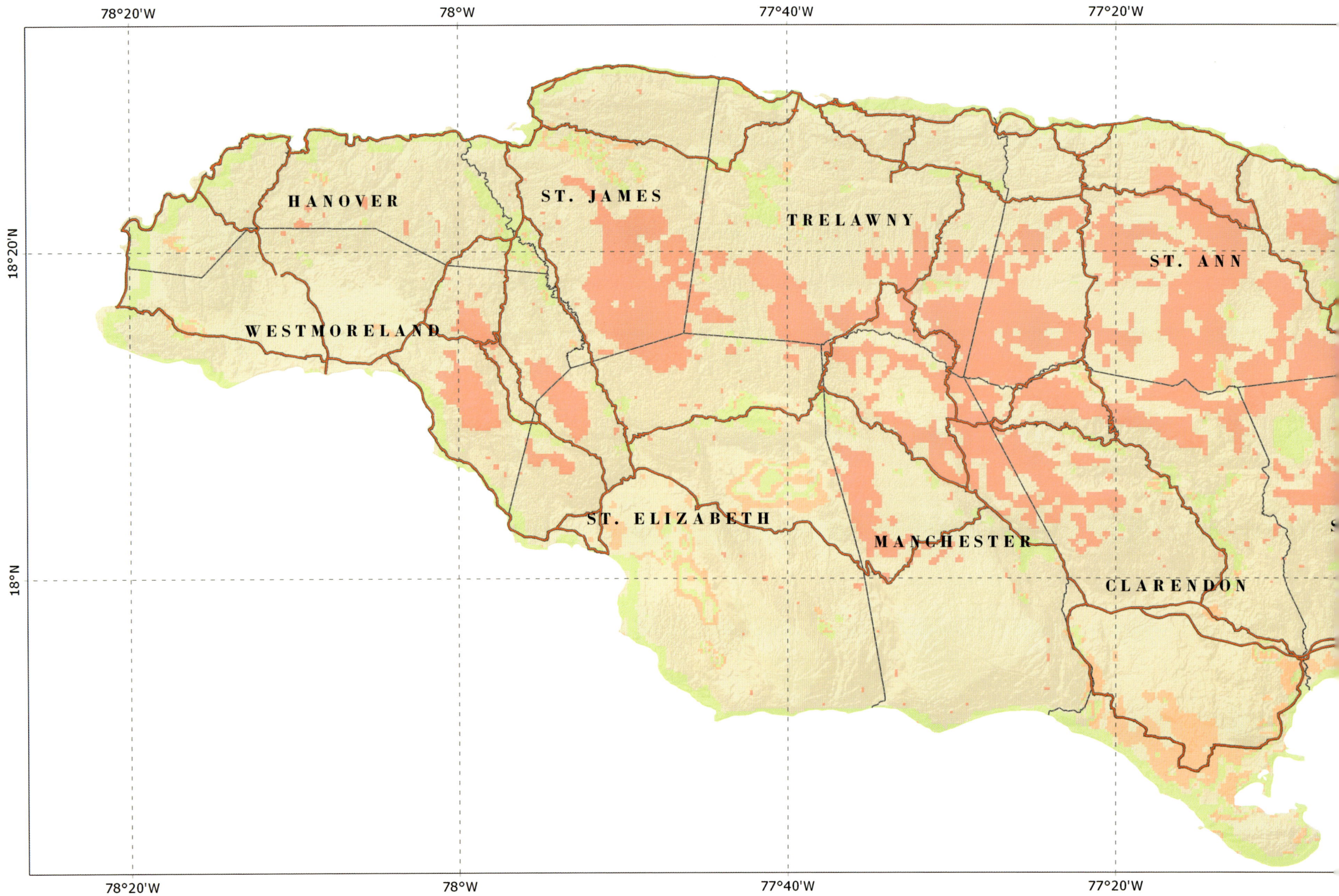

Map 2.2.2d Wind speed, 100-year return period

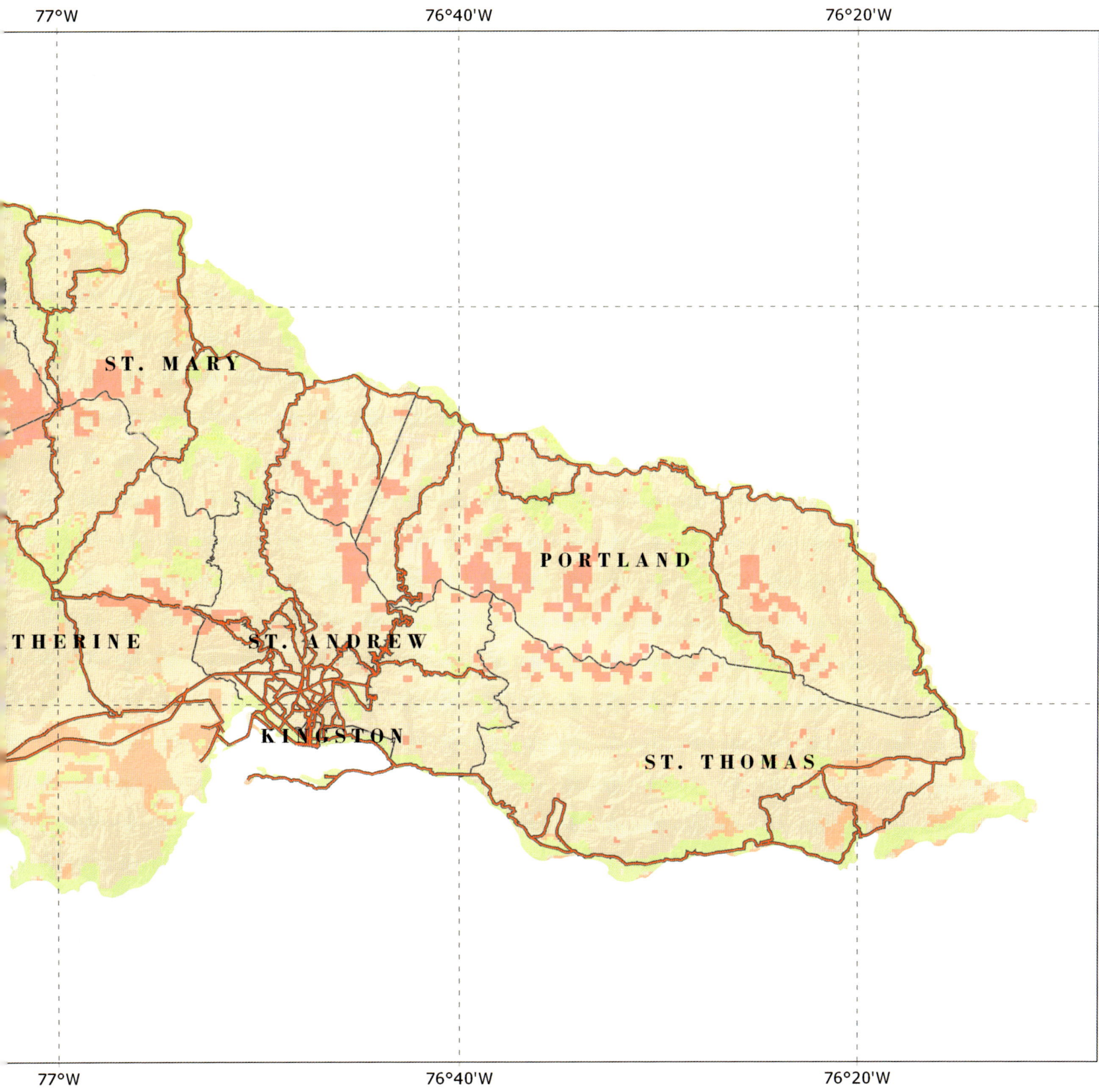

100-year return model

Wind speed (mph)

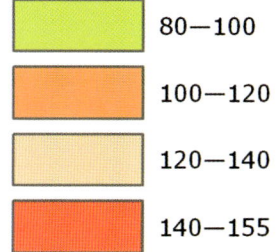

- 80—100
- 100—120
- 120—140
- 140—155

⸺ Parish boundary

⸺ Main road

ST. MARY

THERINE

ST. ANDREW

KINGSTON

PORTLAND

ST. THOMAS

77°W
76°40'W
76°20'W

18°20'N

18°N

1:450,000

| 0 | 5 | 10 | 20 kilometres |

| 0 | 5 | 10 | 20 miles |

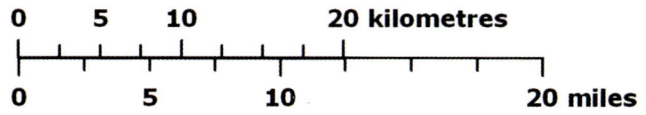

Map Datum:
Latitude-Longitude geodetic grid
World Geodetic System 1984 datum
Prime Meridian: Greenwich
Angular unit: Degree

Plate 2.2.5 Damage resulting from Hurricane Dean, 2007 (P. Lyew-Ayee Jr, 2007)

2.3 COASTAL FLOODING

L ow-lying areas along the coast are vulnerable to flooding from storm surges associated with tropical storms and hurricanes, or from tsunamis caused by earthquakes. Throughout Jamaica's recorded history, there have been spectacular examples of coastal flooding, ranging from the tsunami that accompanied the 1692 earthquake, to the recent storm surges associated with Hurricane Ivan and Hurricane Dean in the first decade of the twenty-first century.

As an island nation that relies heavily on a tourism economy largely centred on beaches, cruise shipping and other coastal attractions, special attention and care have to be given when developing infrastructure in these areas with respect to understanding the nature of coastal flooding.

Table 2.3.2 Saffir-Simpson Hurricane Damage Potential Scale

Storm Category/Wind Speed (kph)	Wave Surge Height (m)
1 (119–154)	1.2–1.5
2 (155–177)	1.8–2.4
3 (178–209)	2.7–3.6
4 (210–249)	4–5.5
5 (>249)	>5.5

Source: Simpson and Riehl 1981.

Table 2.3.1 Vulnerability of Critical Infrastructure to Coastal Flooding in Jamaica

	Total	Police Stations	Fire Stations	Hospitals	Shelters
Number of buildings	504,979	184	33	35	798
Number of buildings below 10 metres above sea level	40,060	29	10	8	43

Table 2.3.3 Lowest-Lying Communities in Jamaica

Rank	Communities (% area below 10 metres above sea level)
1	Passage Fort/Garveymeade, St Catherine
2	Orange Bay, St James
3	Waterford, St Catherine
4	Port Royal, Kingston
5	Edgewater, St Catherine

Plate 2.3.1 Storm surge-prone coastline, Black River, St Elizabeth (R. Ahmad, 2011)

Plate 2.3.2 Displaced rail line at Rocky Point Port, Clarendon, after storm surge associated with the passage of Hurricane Dean (P. Lyew-Ayee Jr, 2007)

Plate 2.3.3 Palisadoes, Kingston, after Hurricane Dean (P. Lyew-Ayee Jr, 2007)

Map 2.3.1 Coastal flooding (classified by elevation)

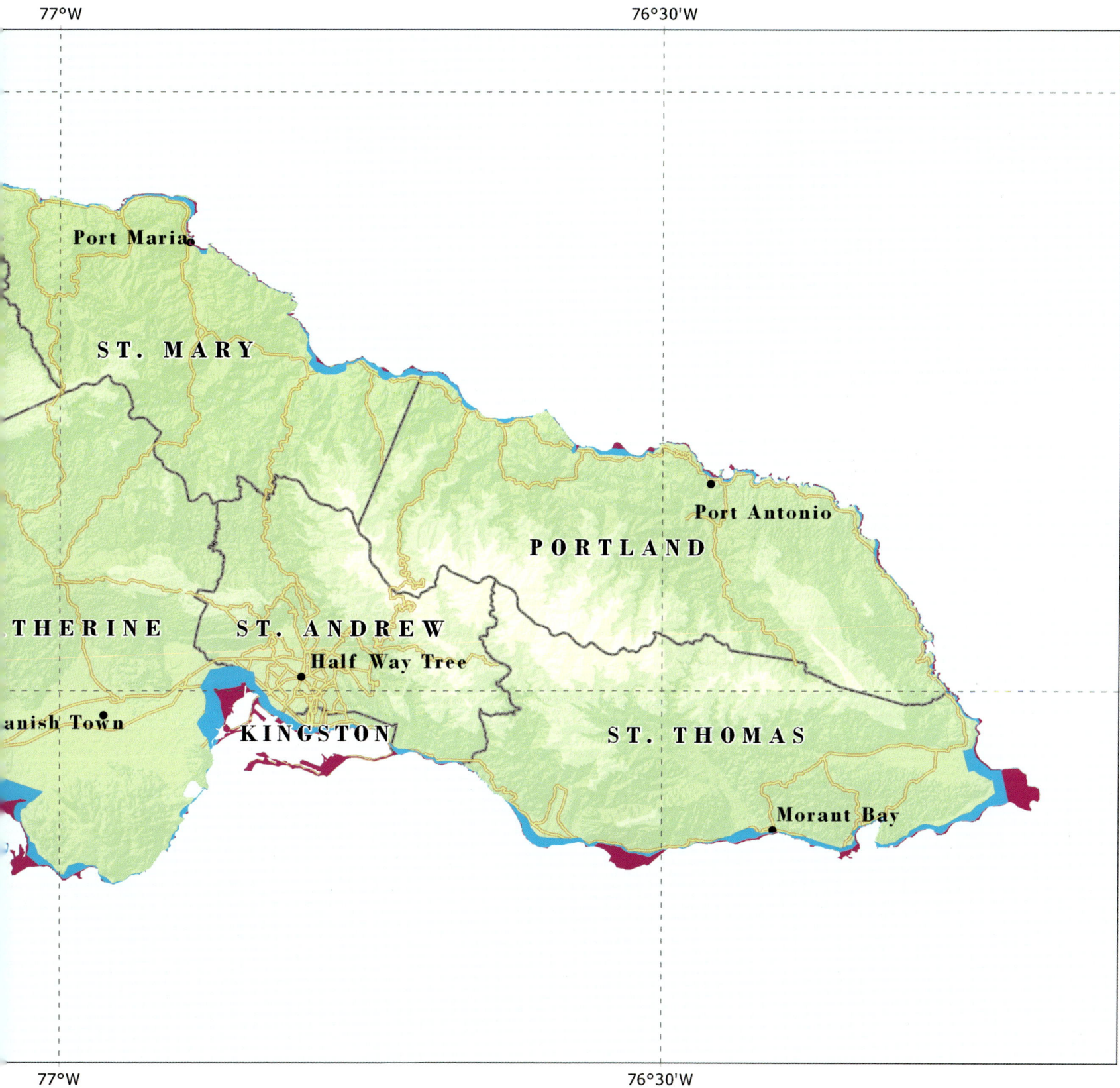

Coastal elevation (metres)

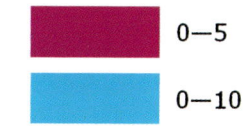

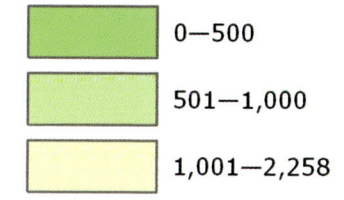

- 0—5
- 0—10
- Parish capital
- Parish boundary
- Main road

Elevation (metres)

- 0—500
- 501—1,000
- 1,001—2,258

77°W 76°30'W 18°30'N

Port Maria

ST. MARY

Port Antonio

PORTLAND

THERINE

ST. ANDREW

Half Way Tree

anish Town

KINGSTON

ST. THOMAS

Morant Bay

18°N

1:450,000

| 0 | 5 | 10 | | 20 kilometres |

| 0 | 5 | 10 | 20 miles |

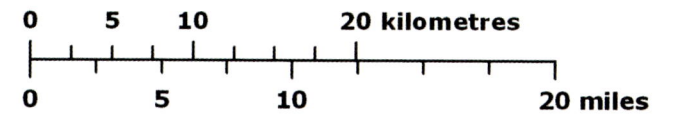

Map Datum:
Latitude-Longitude geodetic grid
World Geodetic System 1984 datum
Prime Meridian: Greenwich
Angular unit: Degree

77°W 76°30'W

Plate 2.3.4 Hurricanes Ivan and Dean, 2004 and 2007, storm surge damage to the Caribbean Terrace area, Harbour View (R. Ahmad, 2007)

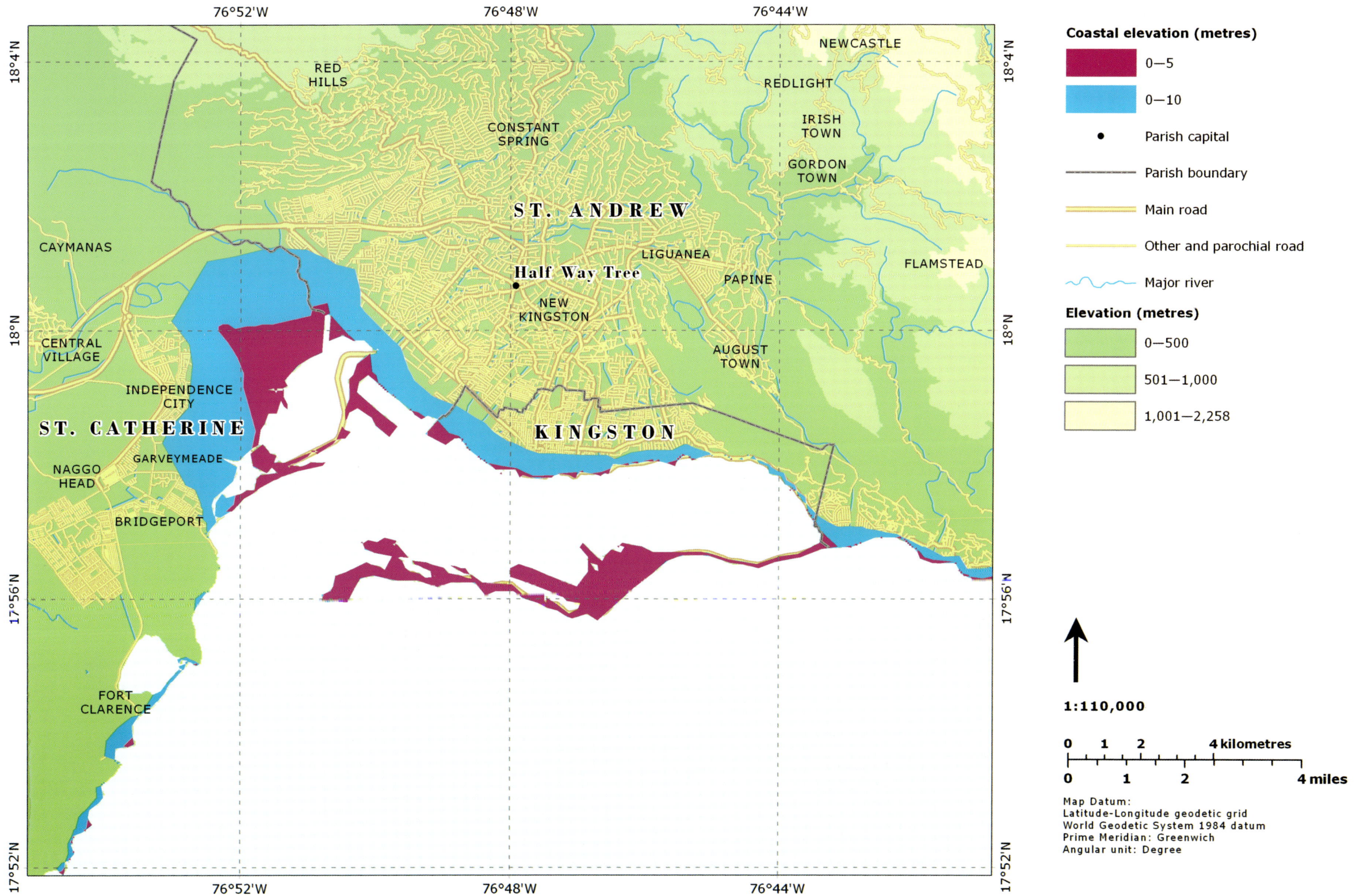

Coastal elevation (metres)

- 0—5
- 0—10
- Parish capital
- Parish boundary
- Main road
- Other and parochial road
- Major river

Elevation (metres)

- 0—500
- 501—1,000
- 1,001—2,258

1:110,000

0 1 2 4 kilometres

0 1 2 4 miles

Map Datum:
Latitude-Longitude geodetic grid
World Geodetic System 1984 datum
Prime Meridian: Greenwich
Angular unit: Degree

Map 2.3.2 Coastal fooding (classified by elevation) of Kingston Metropolitan Region

2.4 INLAND FLOODING

Inland floods are the result of a combination of hydrological processes, generally resulting from a rainfall trigger event that exceeds the threshold of the environment to accommodate the increased water input, and topographical controls. There are typically two types of floods: water floods and debris floods.

Floods usually occur along inland alluvial plains. Flooding is also exacerbated by urbanization, with increasing areas of paved surfaces and deforestation. Flooding along major river valleys is also prevalent, as are cases of karst flooding, especially in Moneague and Cave Valley in St Ann, Harmons and Porus in Manchester, and New Market in St Elizabeth.

Floods include both water and debris flooding, the latter being caused by debris entrained in water and moving fluidly, much like a cement slurry. Both types of flood can inundate and cover homes and infrastructure, and make areas impassable. Following the effects of gravity, they pool and collect in areas where run-off is impeded.

Floods may be the secondary effect of another hazard, such as hurricanes, but also landslides, where major slides obstruct run-off, leading to flooding upstream. Flooding can also be entirely man-made – dam failures or broken water mains are some common examples.

Heavy rainfall may cause flooding when precipitation exceeds the ground's ability to allow for infiltration. High levels of run-off into rivers may lead to these over-topping their banks and flowing over. Excess overland flows not associated with rivers can also lead to flooding if these are not diverted or channelled properly. Blocked drains are common reasons for flooding within residential communities which are otherwise engineered to carry run-off away from these developments.

Historically, from the newspaper archives, St Catherine, followed by St Mary, has had the most reported floods. The largely limestone parish of Manchester has the fewest reported floods over the period.

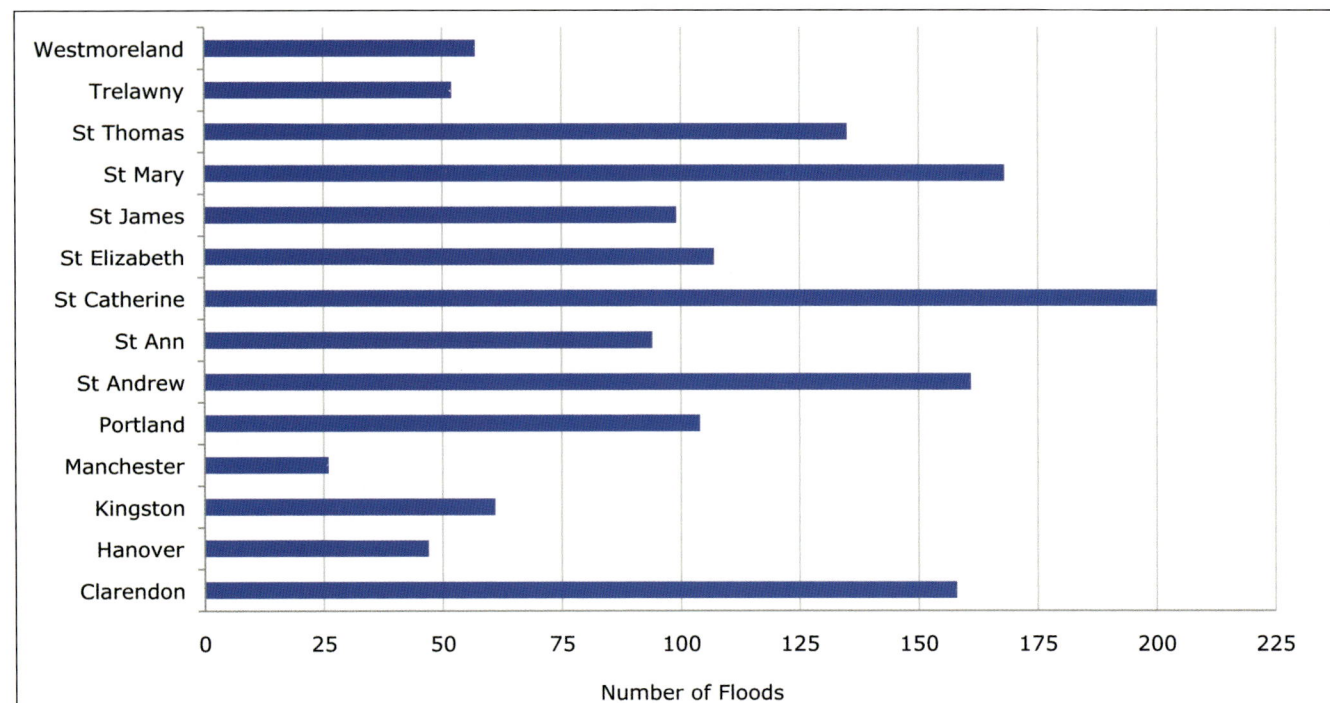

Graph 2.4.1 Floods by parish, 1834–2008

Plate 2.4.1 Flooding in Moneague, St Ann (P. Lyew-Ayee Jr, 2005)

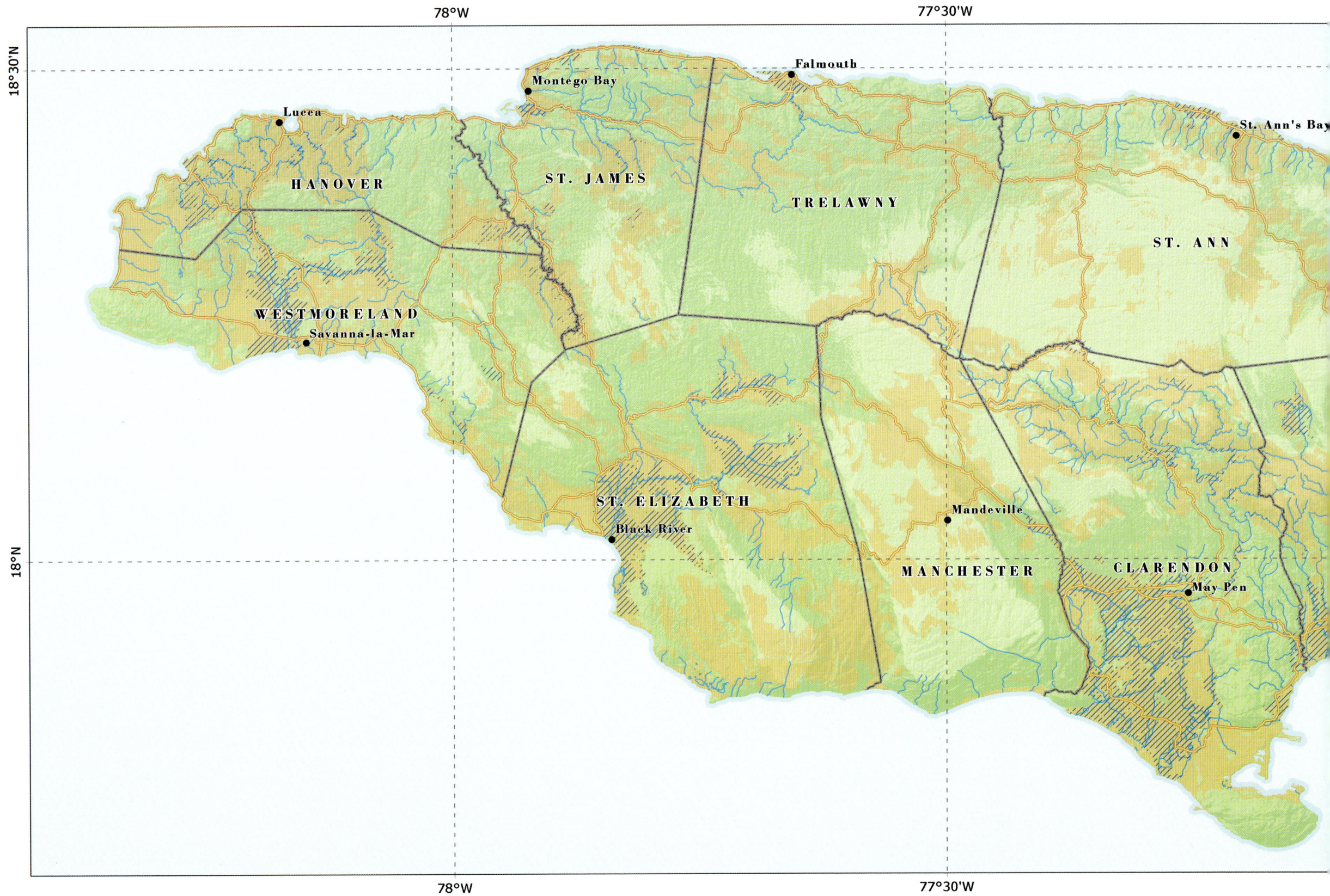

Map 2.4.1 Inland fluvial and debris flooding model of susceptible areas

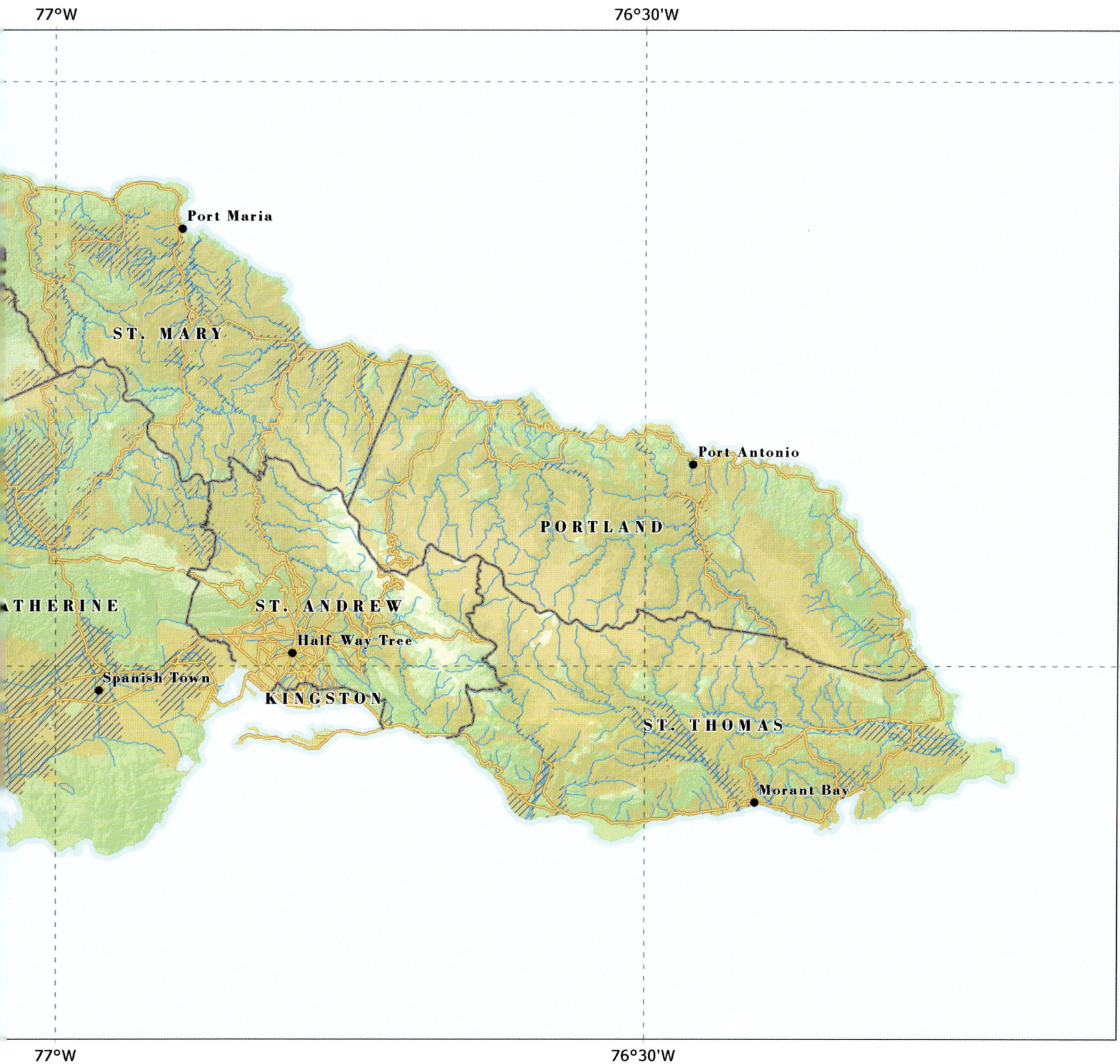

Port Maria

ST. MARY

Port Antonio

PORTLAND

ATHERINE

ST. ANDREW

Half Way Tree

Spanish Town

KINGSTON

ST. THOMAS

Morant Bay

18°N

Fluvial flood

Debris flooding

Parish capital

Parish boundary

Main road

Major river

Elevation (metres)

0—500

501—1,000

1,001—2,258

1:450,000

0   5   10   20 kilometres

0   5   10   20 miles

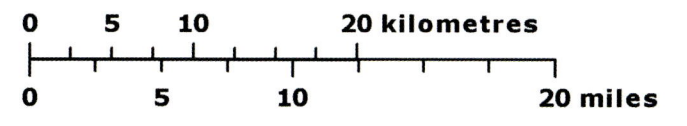

Map Datum:
Latitude-Longitude geodetic grid
World Geodetic System 1984 datum
Prime Meridian: Greenwich
Angular unit: Degree

Plate 2.4.2 Flooding in Kennedy Grove, Alley, Clarendon (R. Ahmad, 2007)

2.5 LANDSLIDES

Landslides typically occur in upland areas with steep slopes and relatively weak bedrock. These are common in non-limestone bedrock areas throughout the island. However, spectacular landslide landforms are also known in limestone country, where bedrock is heavily fractured and structurally weak. These factors, combined with increased surface run-off, promote debris flows, mudslides and debris avalanches. Landslides can fundamentally alter a landscape's shape and form.

Development in landslide-prone areas should consider the structural integrity of the bedrock as well as the broader geological system in which landslides operate. Historically, from the newspaper archives, St Mary has had the most reported landslides, followed by Portland. The parish of Kingston has the fewest.

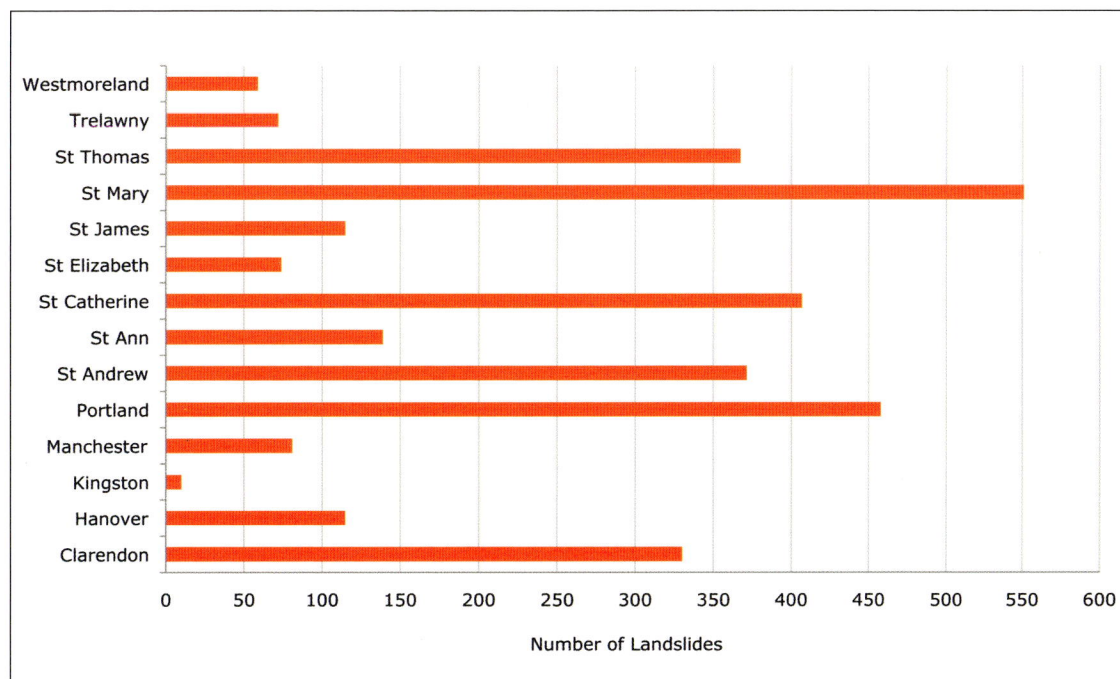

Figure 2.5.1 Graphical illustration of mass movement

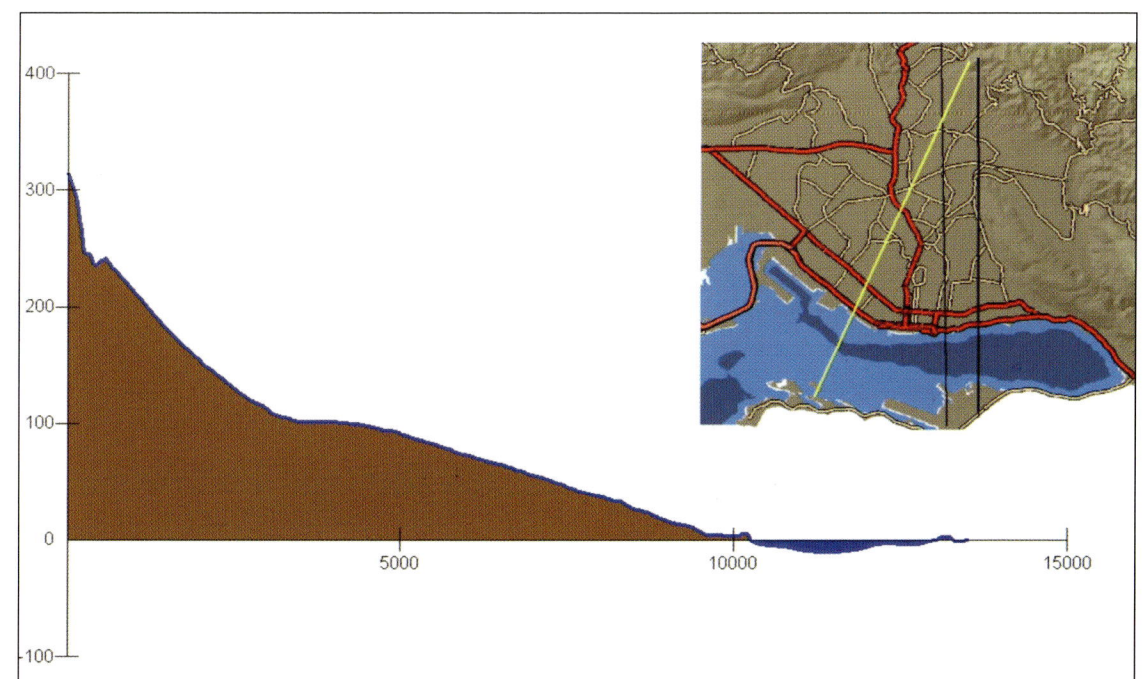

Graph 2.5.1 Landslides by parish, 1834–2008

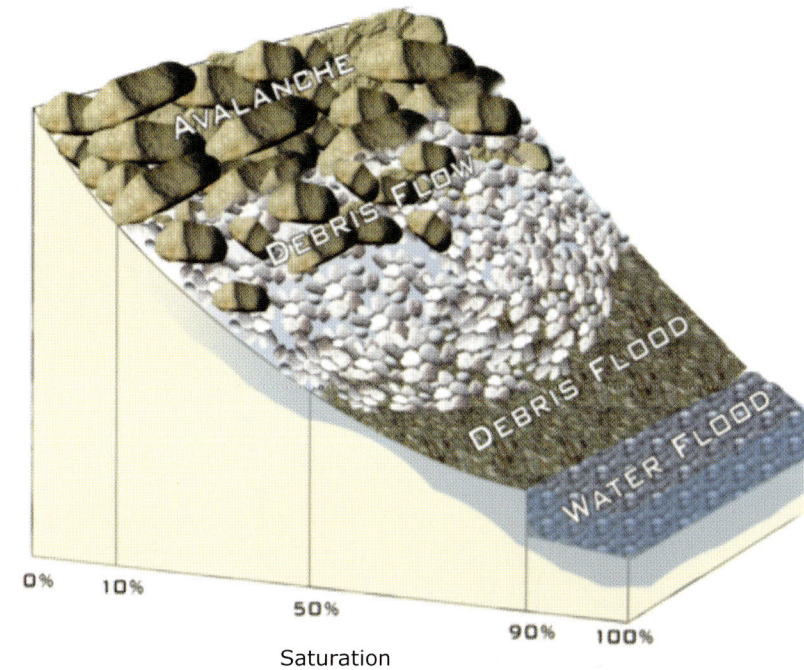

Figure 2.5.2 Cross-sectional view of Kingston

Plate 2.5.1 Hurricane Ivan rainfall-induced landslides, near Guava Ridge, St Andrew (R. Ahmad, 2004)

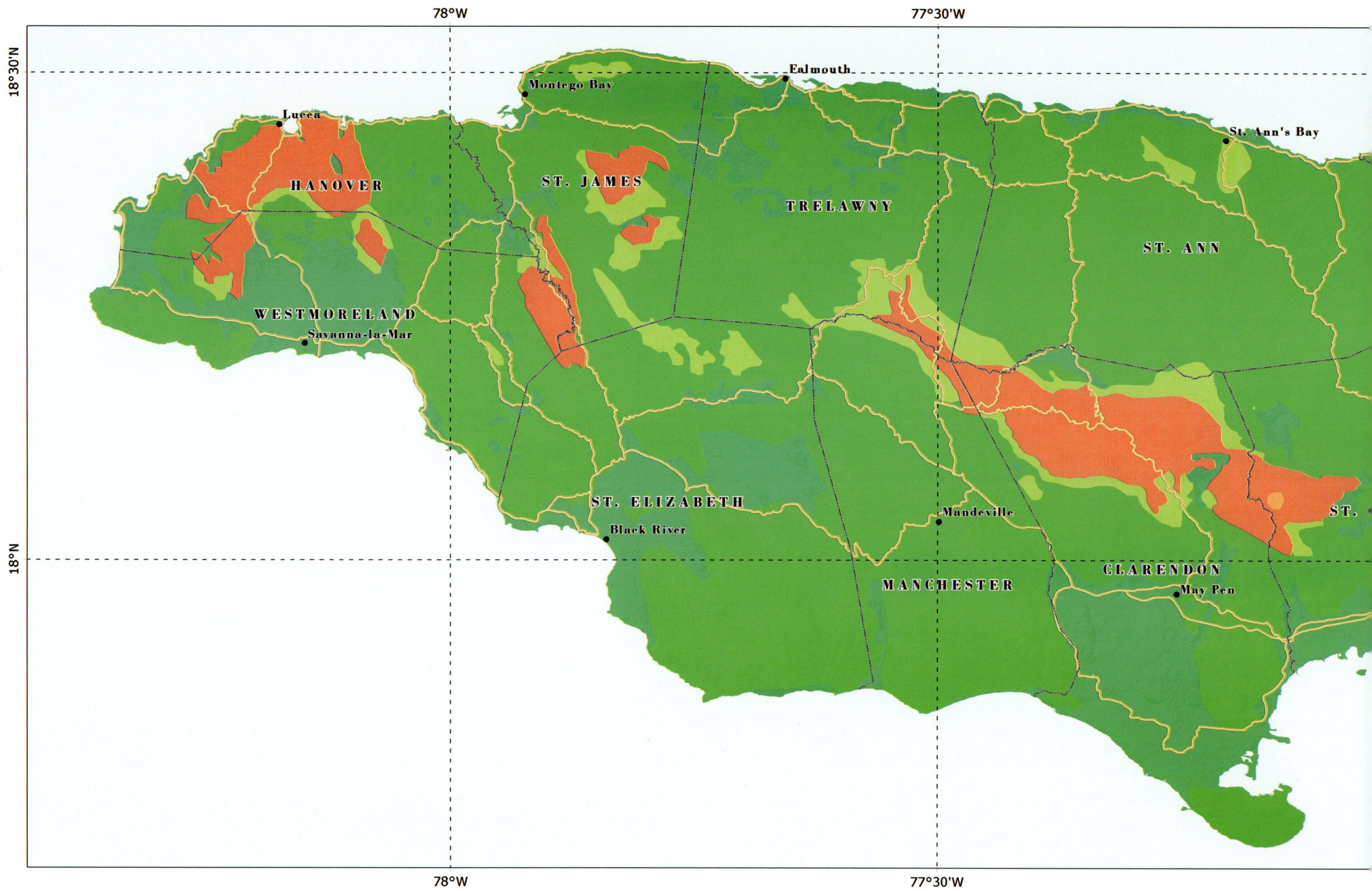

Map 2.5.1 Landslide engineering effort

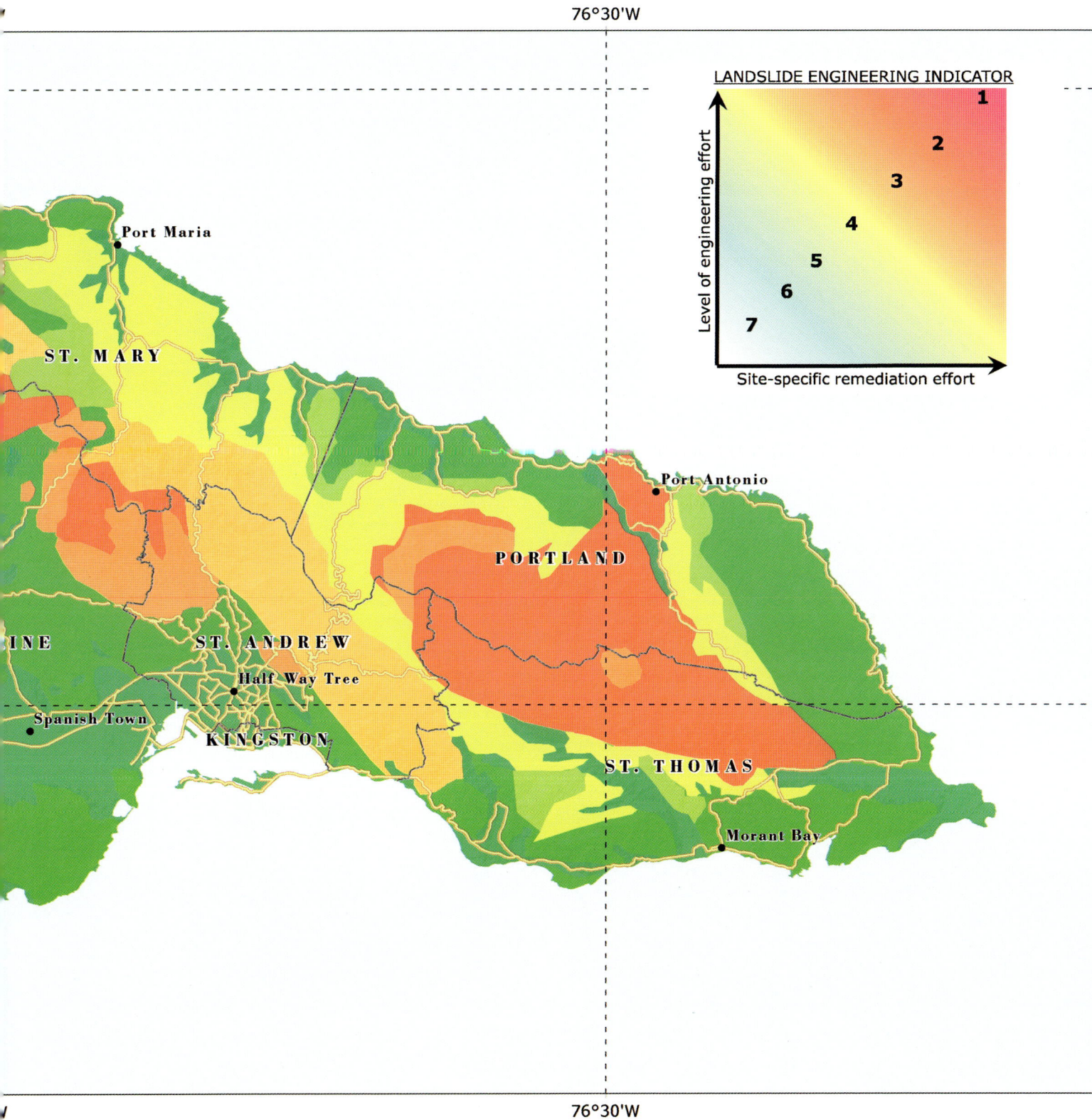

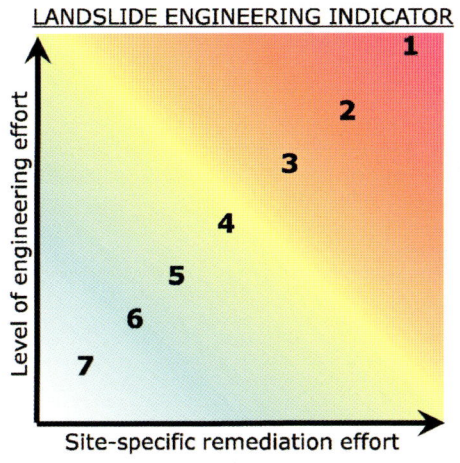

Landslide engineering class

Engineering recommendations

LANDSLIDE ENGINEERING INDICATOR

Level of engineering effort

Site-specific remediation effort

1
2
3
4
5
6
7

1 Highest effort/Highly refined site-specific remediation:
considered active zones in which any development should be avoided; requires improvement in drainage; implement slope vegetation; employ appropriate landslide controls for rotational and translational slides.

2 High effort/Highly refined site-specific remediation:
considered active zones subject to debris-flow; prone to severe landslides and erosion in which new development should be avoided; requires improvement in drainage; implement retaining walls, slope cribworks and anchoring.

3 Medium-High effort/Refined site-specific remediation:
considered active zones subject to multiple hazards; implement appropriate rock slope engineering measures for complex slides.

4 Medium effort/Refined site-specific remediation:
considered active zones requiring the improvement of surface and sub-surface drainage; requires a detailed study for countermeasures of complex slides.

5 Medium-Low effort/Standard site-specific remediation:
considered active zones for lateral spreads; avoid control works for topographic considerations as these may be expensive and difficult.

6 Low effort/Generalized remediation:
zones subject to individual rock falls; implement appropriate rock slope engineering measures.

7 Lowest effort/Generalized remediation:
zones subject to creep and subsidence, requiring improvement in drainage; implement slope protection, piling and anchoring for fills; employ appropriate design criteria for liquefaction and surficial slide effects.

● Parish capital

Parish boundary

Main road

1:450,000

0   5   10   20 kilometres

0   5   10   20 miles

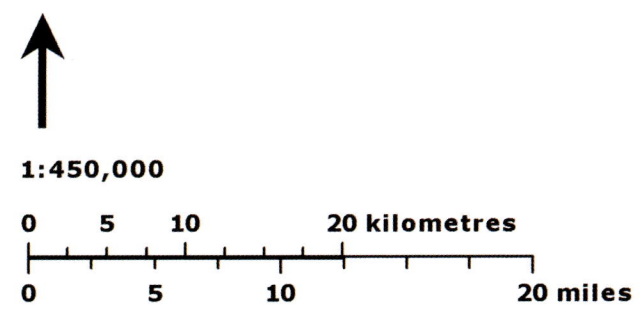

Map Datum:
Latitude-Longitude geodetic grid
World Geodetic System 1984 datum
Prime Meridian: Greenwich
Angular unit: Degree

Plate 2.5.3 Landslide along Jack's Hill Road, St Andrew, after Hurricane Dean (P. Lyew-Ayee Jr, 2007)

Plate 2.5.4 Landslide in a residential area of Kingston (R. Ahmad, 1993)

Table 2.5.1 Historical Landslide Impact in Jamaica

Event and Comments	Landslide Type	Number of Persons Killed/ Injured/Missing	Economic Cost
1. June 7, 1692, earthquake Recorded event along Bog Walk Gorge, St Catherine. Accompanied by extensive deforestation and erosion; specific data on landslides not available	Widespread landslides island-wide; liquefaction and submarine landslides, rock falls and avalanches, and other types of landslides; landslides dammed rivers	Not available	Not available
2. Judgement Cliff landslide, induced by rainfall of October 1692 (1693?) Accompanied by deforestation.	Complex rockslide-slump; approximate volume 6.6×10^6 m³; Yallahs River dammed, caused flooding	At least 19 killed	Homes and plantations destroyed
3. January 14, 1907, earthquake Accompanied by deforestation and erosion; specific data on landslides not available	Widespread landslide activity in eastern Jamaica; liquefaction and submarine landslides; landslide dams	3 killed	Not available
Whitfield Hall landslide, Cascade River, Portland; torrential rains of 1909 Extensive deforestation and erosion	Complex landslide	None	Hundreds of hectares of cultivated land and coffee estates destroyed
5. Millbank landslide, Rio Grande, Portland, November 24, 1937, torrential rains Accompanied by deforestation and erosion	Complex rockslide-slump; volume 2×10^6 m³; dammed the Rio Grande river for six months; caused flooding	33 killed with landslide dam breach; missing and injured not known	Extensive damage to agricultural land, livestock, housing; road blocked, two bridges destroyed
6. Chelsea landslide, Swift River, Portland, November 1940; torrential rains Accompanied by deforestation and erosion	Complex rockslide-slump; Swift River dammed; caused flooding	10 killed – landslide dam breached	Extensive damage to agricultural property, roads, bridges
7. March 1, 1957, earthquake	Rock falls, rock slides, debris slides mainly in western part of the island	2 killed, 7 injured	Railway line blocked, damage statistics not available
8. Hurricane Flora, October 8, 1963 Accompanied by deforestation and erosion	Extensive debris flows islandwide; Mahogany Vale footbridge landslide (volume 40,000 m³) dammed Yallahs River; cable breaks suggestive of submarine landslides near the mouths of Yallahs and Hope Rivers	None	Agriculture, houses and roads damaged island-wide; damage statistics not available
9. June 10, 1969, heavy rainfall event	Extensive debris flows island-wide	None	Extensive damage to public and private property; damage statistics not available

Table 2.5.1 Historical Landslide Impact in Jamaica (*cont'd*)

Event and Comments	Landslide Type	Number of Persons Killed/Injured/Missing	Economic Cost
10. June 12, 1979, tropical depression–associated rainfall event Dropped about 865mm of rain in 8–10 hours in western Jamaica; approximately 160,000 people across the region were affected; disruption of communication, scouring of roads, houses swept away; accompanied by deforestation and erosion	Extensive debris flows and sediment water flows in some 2,500 km^2 in the western parishes	44 persons killed	Total damage to housing and infrastructure in excess of $70 million; houses, roads and agricultural fields buried under landslide debris
11. Hurricane Frederick rainfall event, September 11–13, 1979	Extensive debris flow activity in eastern Jamaica	None	Extensive damage; data not available
12. May–June 1986, flood rains Accompanied by deforestation and erosion	Extensive debris flow activity in eastern and central Jamaica	None	Roads and bridges damaged and destroyed
13. Preston Landslide, St Mary, March 1986	Slide and lateral spread, area 100ha, volume 11.9 x 106 m^3	None	Village of Preston destroyed; 17 families displaced; replacement cost in 1986 $273,000
14. Hurricane Gilbert rainfall event, September 12, 1988 Approximately 60 per cent of island's water facilities damaged; repairs to island's road network at $19.3 million; landslides delivered approximately 20,000m^3 of sediment to rivers; accompanied by deforestation and erosion	Extensive debris flows in interior mountain ranges of eastern and central Jamaica	None	Extensive damage; 478 landslides along 108 km of roads in the Above Rocks area of St Andrew; Boar River water supply pipeline damaged; landslide damage approximately $25 million
15. May 21–22, 1991, flood rains Accompanied by deforestation and erosion	Extensive debris flows in eastern and central parishes	1 killed	Extensive damage to agriculture, roads, bridges, private and public property; Bog Walk Gorge road blocked by a landslide for 6 months
16. January 13, 1993, earthquake, Kingston and St Andrew	Rock falls, rock slides, debris slides in eastern Jamaica; cable breaks off the coast of Kingston suggesting submarine landslides	1 killed	Not available
17. Tropical Storm Gordon, November 11–12, 1994 Accompanied by deforestation and erosion	Mostly debris flows and mudflows in eastern and central Jamaica	None	Approximately 241 km of island's total road network damaged; total damage $2 million; damage to water systems at $834,000

Table 2.5.1 continues

Table 2.5.1 Historical Landslide Impact in Jamaica (*cont'd*)

Event and Comments	Landslide Type	Number of Persons Killed/ Injured/Missing	Economic Cost
18. January 3–4, 1998, flood rains, Rio Grande Valley, Portland Accompanied by deforestation and erosion	Mostly debris flows and mud flows	4 persons buried under landslide debris	Landslide damage $4.7 million (60 per cent of total damage at $7.84 million); landslide damage to agriculture $1.4 million, water systems $169,000, roads and houses $3.08 million
19. Prehistoric landslides west of Ewarton and Mt Diablo, St Catherine	Tertiary limestone slipped over granodiorite and volcaniclastic sediments, Cretaceous basement as complex slides and flows	Not available	Not available
20. Prehistoric Rio Nuevo Valley landslides, Guys Hill–Pembroke Hall–Lambkin Hill area, St Mary	Lateral spreads and rock slump flow; caps of Tertiary limestone and sandstone slipped over mudrocks	Not available	Not available
21. Prehistoric landslide, Williamsfield, St Catherine	Debris–rock slide in grandiorite	Not available	Not available
22. Prehistoric landslides along the Liguanea Ridge and Stony Hill in St Andrew	Large landslides along the Wagwater Fault Zone; complex slide	Not available	Not available
23. Spur Tree Fault Zone landslides, Manchester	Tertiary limestone slipped over weak basement rocks as complex rock–debris slides and flows	Not available	Not available
24. Prehistoric landslides in the upper Hope River valley, Enfield, near Gordon Town, St Andrew	Rock slides and rock avalanches; landslide materials blocked the Hope River and created a temporary lake	Not available	Not available
25. Prehistoric Burlington landslide, Rio Grande Valley, Portland	Complex rock slide and flow, area 40 ha, approximate volume 14 million m^3; landslide debris blocked the Rio Grande and created a lake that extended some 8 km upstream	Not available	Not available
26. Prehistoric Jupiter landslide, Rio Grande Valley, Portland	Complex rock slide and flow, landslide area 3.0 km^2; limestone cap rocks slipped over shales; landslide debris blocked Rio Grande and created a lake	Not available	Not available

2.6 HISTORICAL EARTHQUAKES

Earthquakes are ground-shaking events that occur as a result of the sudden release of energy in the earth's crust. These range in magnitude and intensity across space. Magnitude, which is most familiar when describing earthquakes, refers to the amount of energy released by the earthquake at the epicentre, the geographic location of the earthquake. This is most commonly measured using the Richter scale. The intensity measures the effect the earthquake has on the earth's surface, structures and humans, and may vary depending on location away from the epicentre; there may be several intensity measures for a single earthquake event, while there is only one magnitude per event. Intensity is commonly measured using the Modified Mercalli Intensity Scale. Peak ground acceleration (PGA), measured as a percentage of gravity acceleration, measures earthquake acceleration on the earth's surface and is an extremely important consideration when doing engineering design to withstand earthquakes. All three measures are important when evaluating earthquakes.

Jamaica lies on the northern edge of the Caribbean Plate, near the margin with the North American Plate. While not as tectonically active as the Pacific "Ring of Fire", this region's tectonic history includes the 1692 earthquake which devastated Port Royal, Jamaica; the 1907 earthquake which destroyed Kingston, Jamaica; and the January 2010 earthquake which destroyed Port-au-Prince, Haiti. Its imprint on the earth's surface is also evidenced by the Cayman Trench, to the northwest of Jamaica, the deepest point in the Caribbean Sea, at over 7.6 kilometres deep.

Earthquakes spawn numerous secondary hazards, including tsunamis, landslides (both surface and submarine) and liquefaction. While Jamaica has experienced tsunamis associated with the major events of 1692 and 1907, these were relatively minor compared to those generated by the 2004 Indian Ocean earthquake or the 2011 Japan earthquake. More frequent, however, are the landslides that result from earthquakes, including the spectacular Judgement Cliff in Llandewey, St Thomas, and liquefaction (when, because of ground motion, surface sediments act like a fluid), evidenced by the effect on the former artillery store ("Giddy House") at Fort Charles in Port Royal. Submarine cable breaks also resulted from underwater landslides that occurred as a result of the 1907 earthquake, severing communication lines at the time.

Table 2.6.1 Modified Mercalli Scale and Peak Ground Acceleration

Modified Mercalli Intesity	Peak Ground Acceleration (g)
IV	0.03 and below
V	0.03–0.08
VI	0.08–0.15
VII	0.15–0.25
VIII	0.25–0.45
IX	0.45–0.60
X	0.60–0.80
XI	0.80–0.90
XII	0.90 and above

Source: www.mercallixii.com.

Table 2.6.2 Richter Magnitude, Peak Ground Acceleration and Duration

Richter Magnitude	Peak Ground Acceleration (g)	Duration (seconds)
5	0.09	2
5.5	0.15	6
6	0.22	12
6.5	0.29	18
7	0.37	24
7.5	0.45	30
8	0.5	34
8.5	0.5	37

Source: www.mercallixii.com.

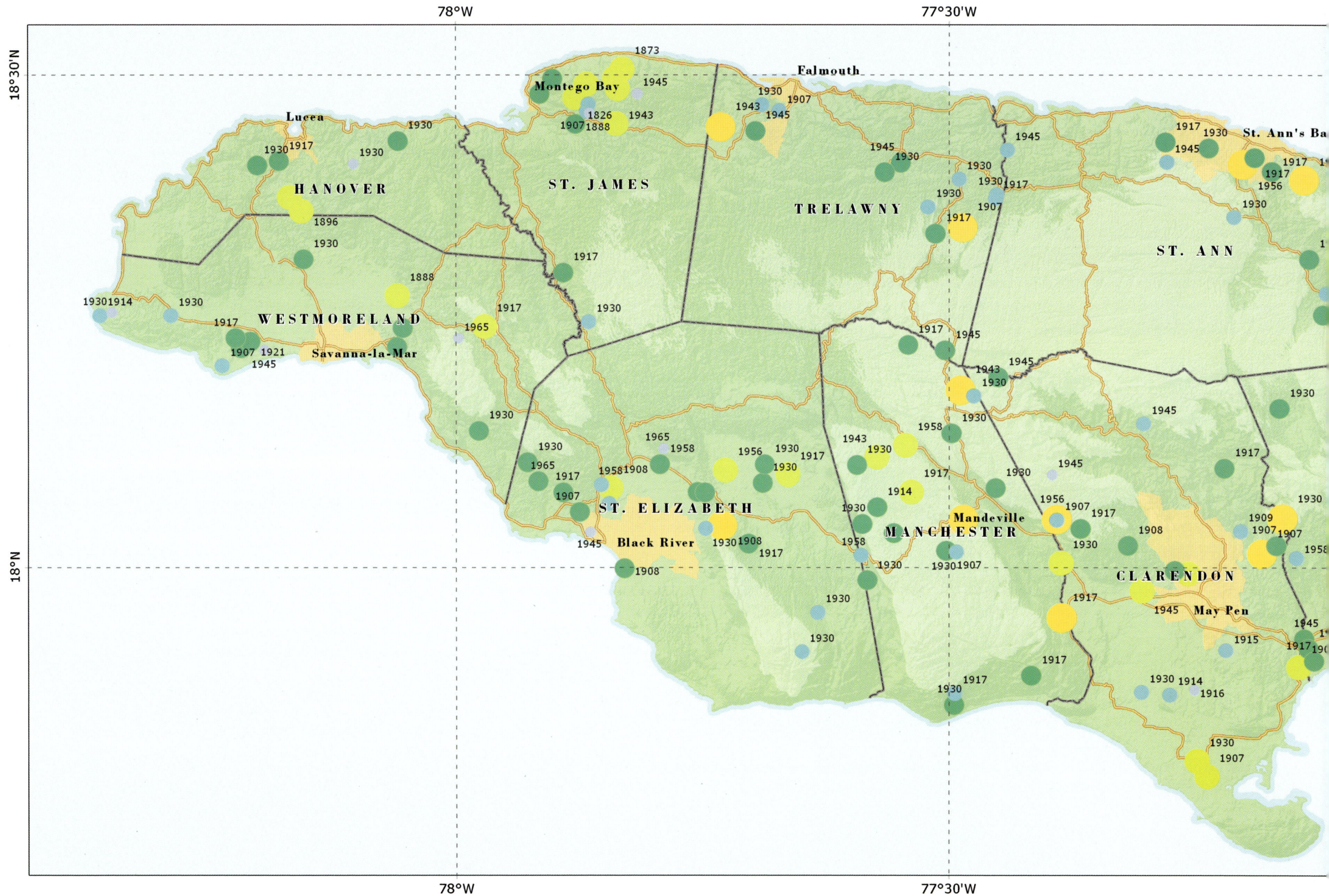

Map 2.6.1 Historical earthquake occurrences of Jamaica

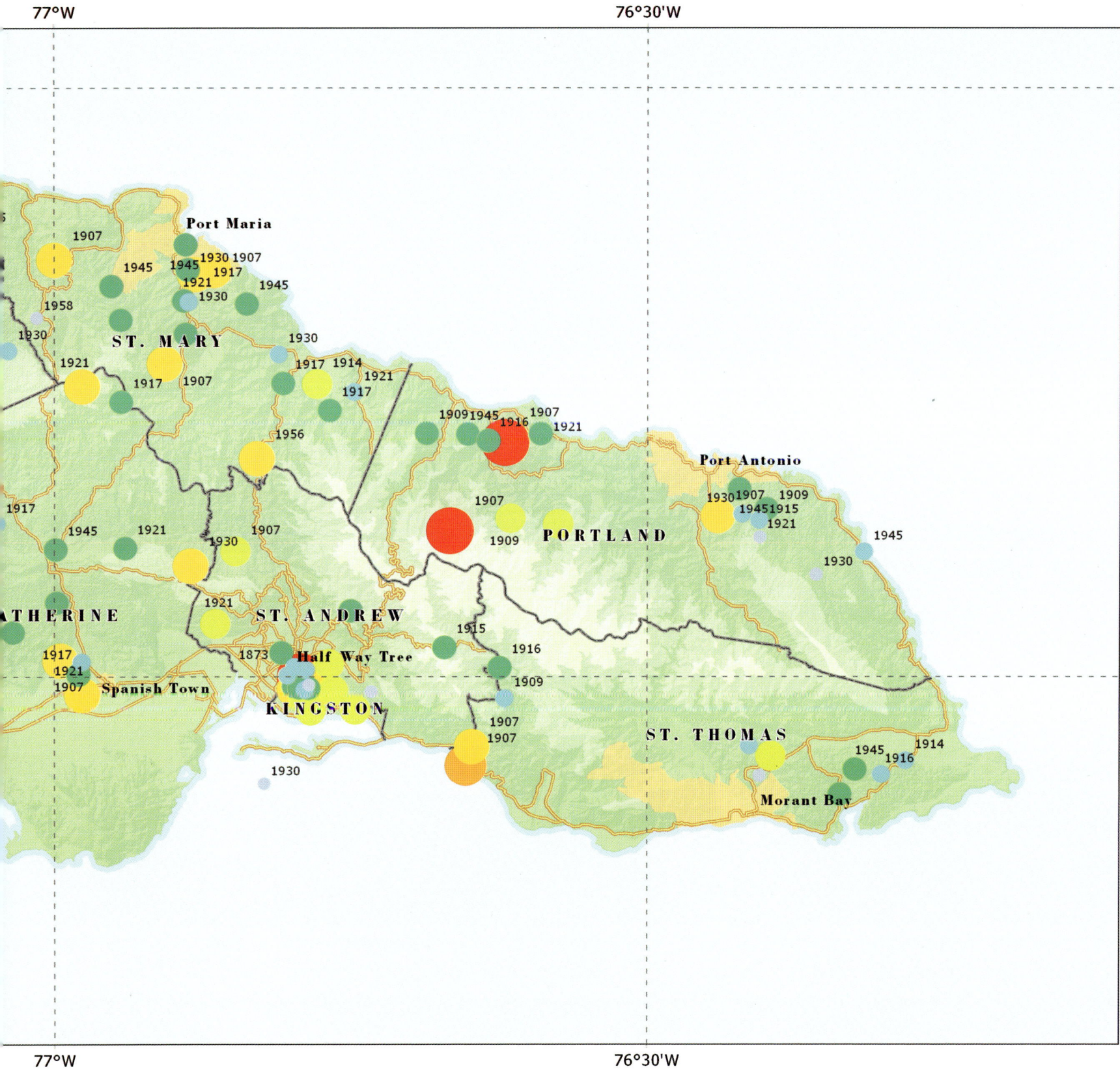

Modified Mercalli Intensity (MMI)

- I—III
- IV
- V
- VI
- VII
- VIII
- IX

Parish capital

Parish boundary

Main road

Elevation (metres)

- 0—500
- 501—1,000
- 1,001—2,258

1:450,000

0 5 10 20 kilometres

0 5 10 20 miles

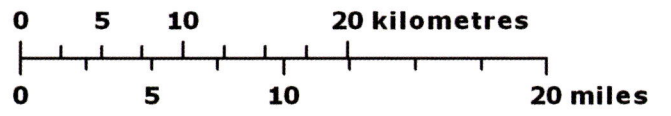

Map Datum:
Latitude-Longitude geodetic grid
World Geodetic System 1984 datum
Prime Meridian: Greenwich
Angular unit: Degree

Plate 3.6.1 Limestone Cliffs, Herring & Thomas (R. Lyew-Ayee Jr. 2006)

THE EARTHQUAKE OF 1907

The January 1907 earthquake which devastated Kingston stands as the second-worst earthquake to have hit Jamaica in recorded history. Although with its epicentre near the St Mary–Portland border, the damage was most acutely felt in Kingston, resulting in catastrophic damage from ground shaking and related effects, as well as from a fire, which exacerbated the disaster situation.

The earthquake measured 6.5 on the Richter scale, occurring around 3:30 p.m. Up to one thousand people died and over ten thousand were made homeless; the population of Jamaica at the time was just over eight hundred thousand. Numerous buildings collapsed, including sections of the Ward Theatre, the Kingston Parish Church, Mico College, and as far north as the Constant Spring Hotel (now the convent at the Immaculate Conception High School).

Plate 2.6.2 Damage to the Coke Chapel, Kingston, as a result of the 1907 earthquake (Institute of Jamaica, 1907)

Plate 2.6.3 Damage to the Mico College, Kingston, as a result of the 1907 earthquake (Institute of Jamaica, 1907)

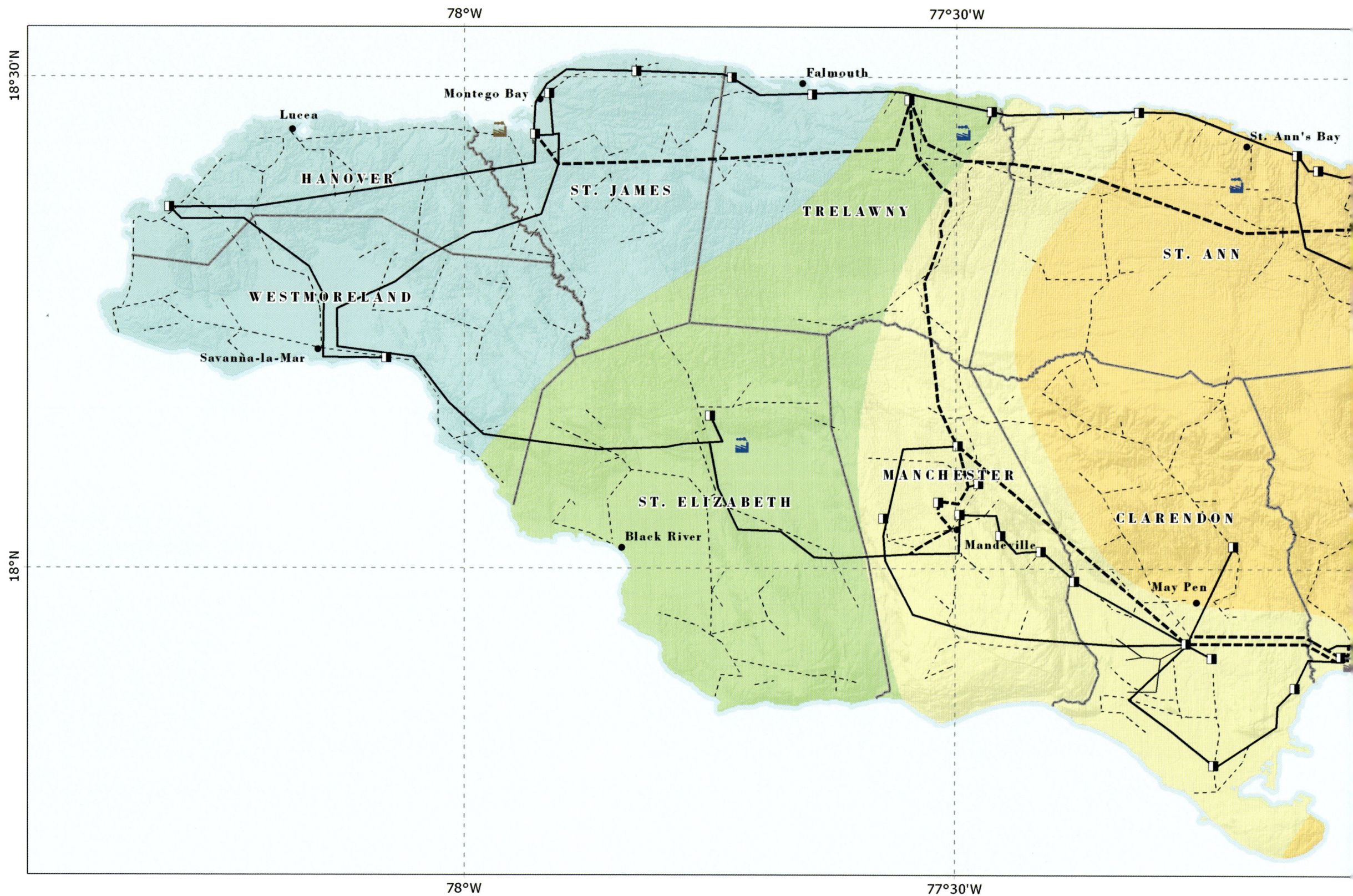

Map 2.6.2 Layout of contemporary information on 1907 earthquake MMI zones

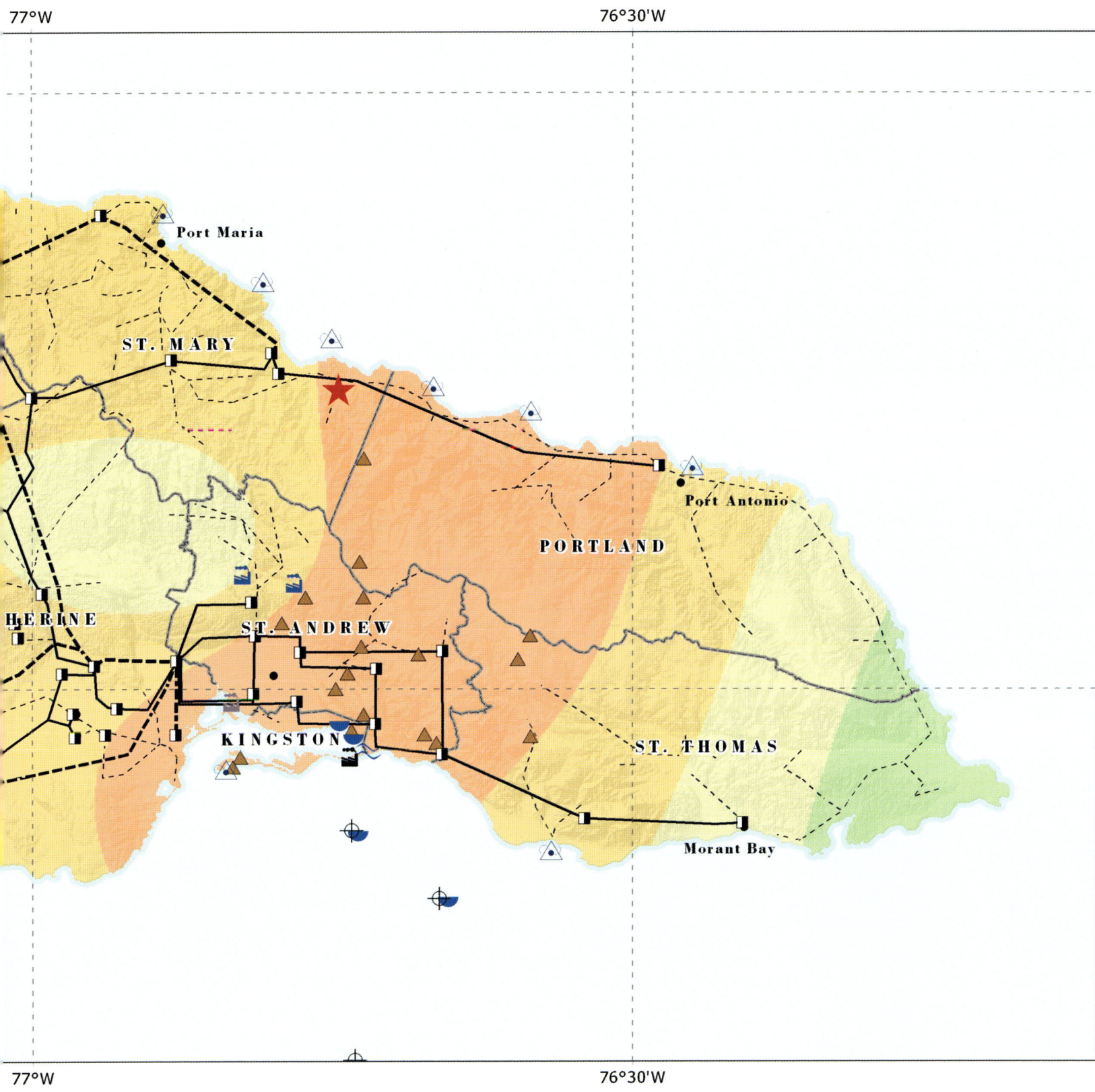

1907 Damage

⭐	Epicentre
⊕	Cable break
🔺	Landslide
≈	Liquefaction
🔵	Submarine slumping
🔺•	Tsunami

Modified Mercalli Intensity (MMI)

🟦	IV—V
🟩	V—VI
🟨	VI—VII
🟧	VII—VIII
🟧	VIII—IX

Transmission line type

– – – – –	138 kV transmission line
————	69 kV transmission line
————	33 kV transmission line
- - - - -	Distribution line

Electricity type

🏭	Diesel
🏭	Hydro
🏭	Steam
🏭	Turbine-gas
▯	Substation
——	Parish boundary
•	Parish capital

1:450,000

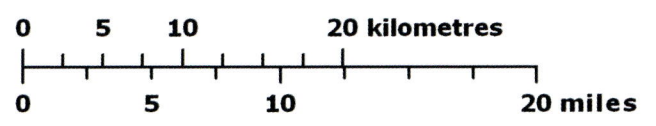

Map Datum:
Latitude-Longitude geodetic grid
World Geodetic System 1984 datum
Prime Meridian: Greenwich
Angular unit: Degree

Table 2.6.3 Detailed Modified Mercalli Intensity Scale

Modified Mercalli Intensity Scale	Shaking	Damage	Description	Modified Mercalli Intensity Scale	Shaking	Damage	Description
I	Not felt	None	Not felt except by a very few under especially favourable circumstances.	VII	Very strong	Moderate	Everybody runs outdoors. Damage negligible in buildings of good design and construction; slight to moderate in well-built ordinary structures; considerable in poorly built or badly designed structures. Some chimneys broken. Noticed by persons driving motor cars.
II	Weak	None	Felt only by a few persons at rest, especially on upper floors of buildings. Delicately suspended objects may swing.	VIII	Severe	Moderate/Heavy	Damage slight in specially designed structures; considerable in ordinary substantial buildings, with partial collapse; great in poorly built structures. Panel walls thrown out of frame structures. Fall of chimneys, factory stacks, columns, monuments and walls. Heavy furniture overturned. Sand and mud ejected in small amounts. Changes in well water. Persons driving motor cars disturbed.
III	Weak	None	Felt quite noticeably indoors, especially on upper floors of buildings, but many people do not recognize it as an earthquake. Standing motor cars may rock slightly; vibration like a passing truck. Duration estimated.	IX	Violent	Heavy	Damage considerable in specially designed structures; well-designed frame structures thrown out of plumb; great in substantial buildings, with partial collapse. Buildings shifted off foundations. Ground cracked conspicuously. Underground pipes broken.
IV	Light	None	During the day felt indoors by many, outdoors by few. At night some awakened. Dishes, windows and doors disturbed; walls make creaking sound. Sensation like a heavy truck striking a building. Standing motorcars rock noticeably.	X	Intense	Extreme	Some well-built wooden structures destroyed; most masonry and frame structures destroyed with foundations; ground badly cracked. Rails bent. Landslides considerable from river banks and steep slopes. Shifted sand and mud. Water splashed over banks.
V	Moderate	Very light	Felt by nearly everyone; many awakened. Some dishes, windows and the like broken; a few instances of cracked plaster; unstable objects overturned. Disturbance of trees, poles and other tall objects sometimes noticed. Pendulum clocks may stop.	XI	Extreme	Few structures remain standing	Few, if any (masonry), structures remain standing. Bridges destroyed. Broad fissures in ground. Underground pipelines completely out of service. Earth slumps and land slips in soft ground. Rails bent greatly.
VI	Strong	Light	Felt by all; many frightened and run outdoors. Some heavy furniture moved; a few instances of fallen plaster or damaged chimneys. Damage slight.	XII	Cataclysmic	Total	Damage total. Waves seen on ground surfaces. Lines of sight and level distorted. Objects thrown upward into the air.

Source: Adapted from Bruce A. Bolt, Abridged Modified Mercalli Intensity Scale, 1993.

PARISH NATURAL HAZARDS PROFILES

Facts and Figures

PARISH NATURAL HAZARDS PROFILES

1. Manchester and St Ann have the least proportion of their land area as low-lying.

2. Kingston and St Andrew (as a single parish), Portland, and St Thomas have the greatest topographic range.

3. Hanover is the lowest-lying parish (with Kingston counted along with St Andrew).

3.1 GEOSCAPE

Each parish has a unique topographic character which is a direct result of the geological and geomorphic controls on the area. Geology, therefore, influences topography, which in turn determines an area's vulnerability to natural hazards.

Manchester and St Ann have a relatively low percentage of their land area below 500 metres above sea level. These parishes are dominantly situated on raised dissected limestone plateaux. Accordingly, these parishes are the least vulnerable to coastal flooding. However, these parishes are vulnerable to karst flooding during times of intense and prolonged rainfall. Cave Valley and Moneague in St Ann, and Porus and Harmons in Manchester have all had recent episodes of significant karst flooding.

With Kingston and St Andrew considered as a single parish in this atlas, Hanover is the low-lying parish, with 100 per cent of its land area at or below 500 metres above sea level.

Kingston and St Andrew, along with Portland and St Thomas, reach 100 per cent of their land area at the highest altitudes (they have the gentlest slope on the hypsometric curve). This is due to the presence of the Blue Mountains in the east. Their topographic character is distinctly different from that of the other parishes. These parishes, owing to their rugged topography, have perennial episodes of landslides and debris flows, where unstable geology interacts with steep terrain in which the effects of gravity are significant. These rugged landscapes also have significant surface run-off, leading to numerous rivers and streams. These rivers and streams often overflow, leading to riverine flooding in these areas.

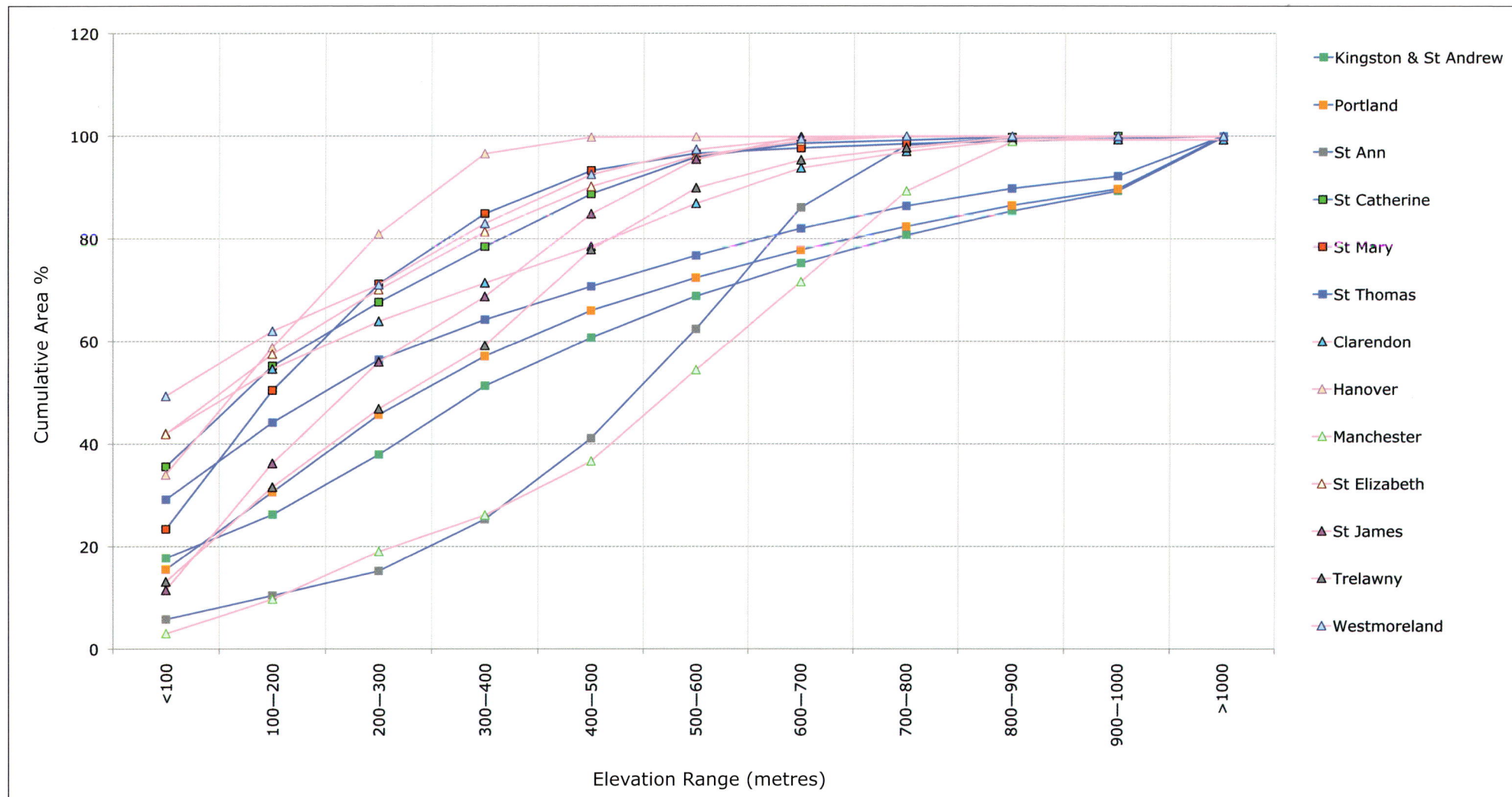

Graph 3.1.1 Hypsometric curve of parishes of Jamaica

3.2 KINGSTON AND ST ANDREW

PARISH PROFILE

Human geography: Kingston and St Andrew together form the administrative capital of Jamaica, with government buildings, legislative and judicial offices, and embassies and high commissions. The population is approximately 600,000 people, and the main economic activities include manufacturing, banking and finance, retail goods and services, and entertainment. Kingston and St Andrew are also the centres for tertiary education and health services for Jamaica. Kingston and St Andrew are administered centrally by the Kingston and St Andrew Corporation (KSAC).

Physical geography: Kingston and St Andrew have highly varied landforms, ranging from the Port Royal, Dallas and Long Mountains in the east, Jack's Hill, Stony Hill, Manning's Hill and Red Hills to the north and northwest, the Liguanea and Harbour View alluvial fans, and the Palisadoes tombolo. Upland areas comprise deeply weathered bedrock, and karst landscape is found in the northwest. The area has high susceptibility to flooding and landslides, as well as with high liquefaction potential along the land-water interface, especially in areas of reclaimed land along the Kingston Waterfront.

Hazard profile: The volcaniclastic Port Royal Mountains and Jack's Hill make up the foothills of the Blue Mountains, which extend from northeast St Andrew into neighbouring Portland and St Thomas. These regions have significant landslide and debris flow activity. Intensive urbanization along the Liguanea Plain has resulted in the paving of surfaces for roadways and buildings, as well as the creation of an artificial gully network to carry run-off. Flooding remains common in the urban areas, largely from poorly planned drainage infrastructure or from blocked drains. There are also examples of landslides and floods caused by non-natural causes, such as broken pipelines and collapsed retaining walls due to poor construction. Coastal regions are exposed to flooding, mostly from storm surges. Kingston and St Andrew have experienced the strongest earthquakes on record, with the 1692 earthquake which devastated Port Royal, and the 1907 earthquake which ravaged Kingston.

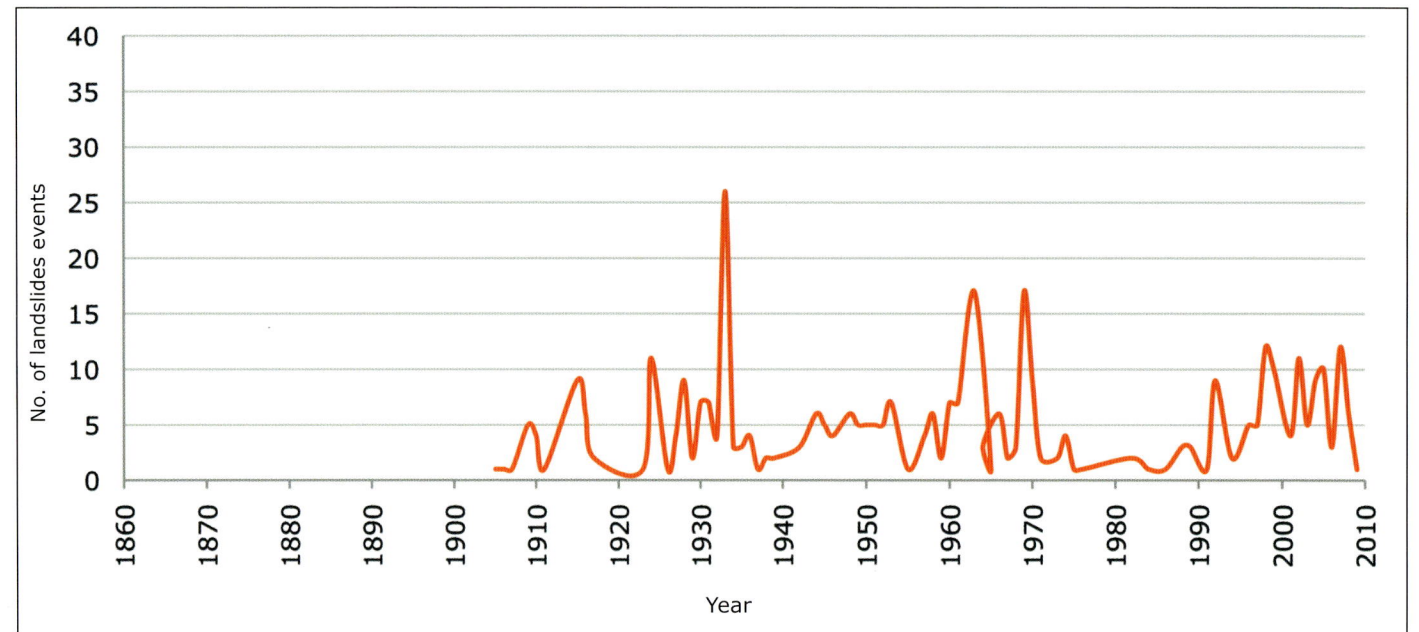

Graph 3.2.1 Kingston and St Andrew landslides

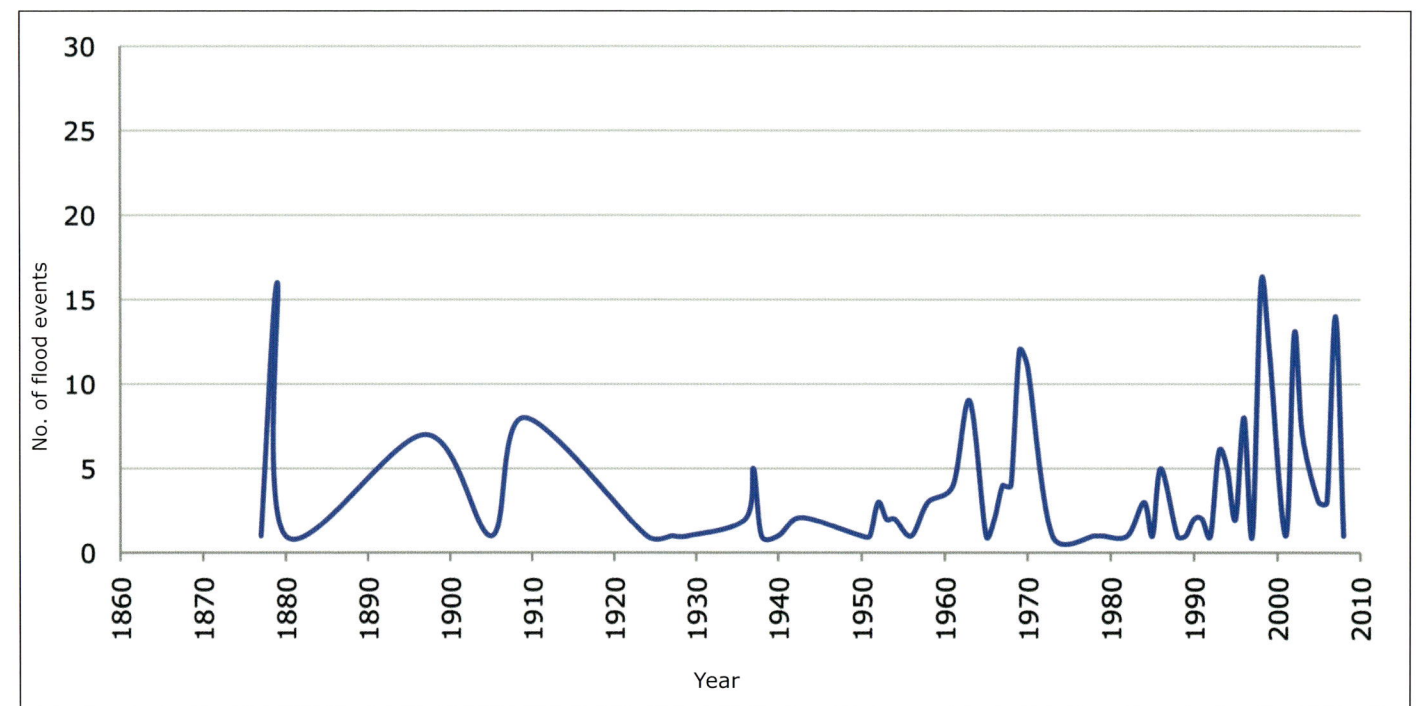

Graph 3.2.2 Kingston and St Andrew floods

Plate 3.2.1 Damage to the Harbour View Bridge as a result of sediment floods following rainfall from Tropical Storm Gustav (P. Lyew-Ayee Jr, 2008)

Plate 3.2.2 Damage to the Caribbean Terrace community from storm surge associated with Hurricane Ivan (R. Ahmad, 2004)

Plate 3.2.3 Rainfall-induced landslide destroyed the road and water supply pipelines during 2004–2008, Skyline Drive, Liguanea Ridge, Upper St Andrew (R. Ahmad, 2008)

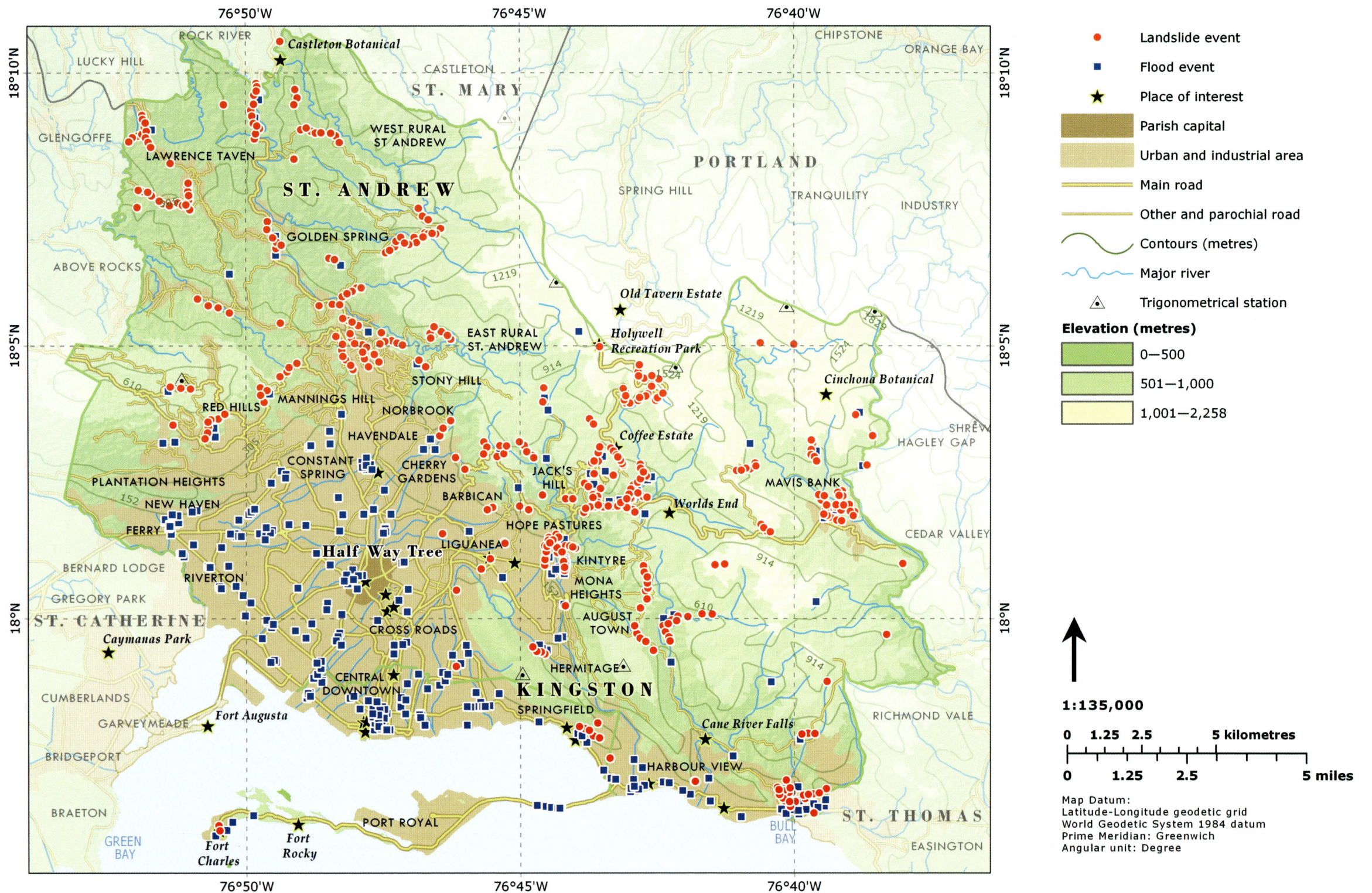

Map 3.2.1 Kingston and St Andrew landslide and flood distribution

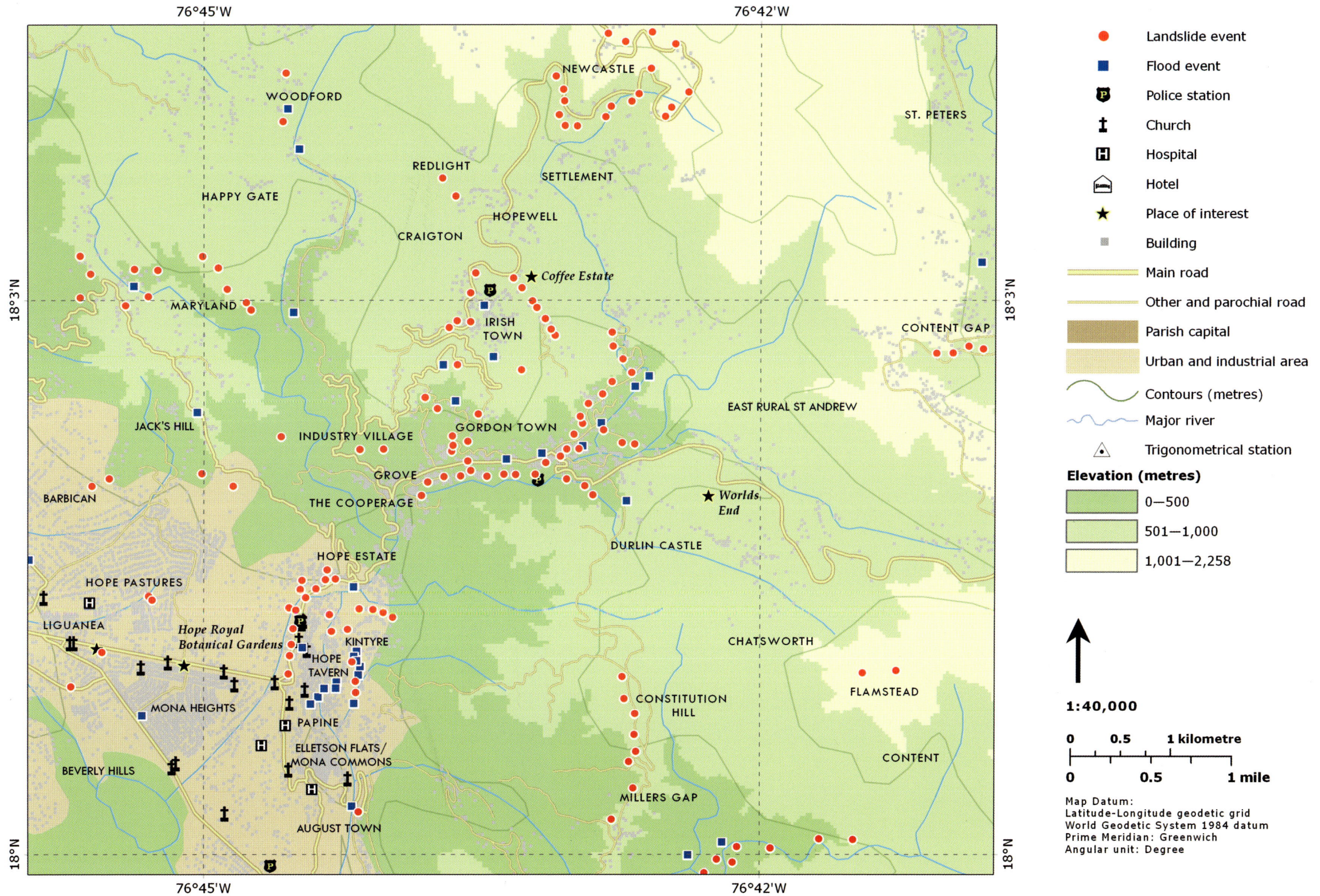

Map 3.2.2 East St Andrew landslide and flood distribution

Plate 3.2.4 Frequent debris flow blocking Highway A3, near Temple Hall, northwest St Andrew, 1991 (R. Ahmad, 1991)

3.3 ST THOMAS

PARISH PROFILE

Human geography: The population of St Thomas is just over 90,000 people. The main economic activity is agriculture, mostly coffee along the slopes of the Blue Mountains, with banana cultivation along the plains and small farming. The capital of St Thomas is Morant Bay.

Physical geography: Southern portions of the parish are made up of alluvial fans of the Yallahs and Morant Rivers, with raised reef platforms also found. Northern parts of the parish are made up of heavily faulted and dissected landscapes in the Blue Mountains. The deeply weathered bedrock is highly susceptible to landslides. The mountain range produces a rain shadow in the parish, resulting in lower-than-average rainfall.

Hazard profile: St Thomas has had many incidents of damage to property and infrastructure from hazard impacts. Riverine flooding has destroyed many bridges and isolated numerous communities, while landslides further upslope have also cut off communities. Hurricanes have also had significant impacts, especially on agricultural activities in the

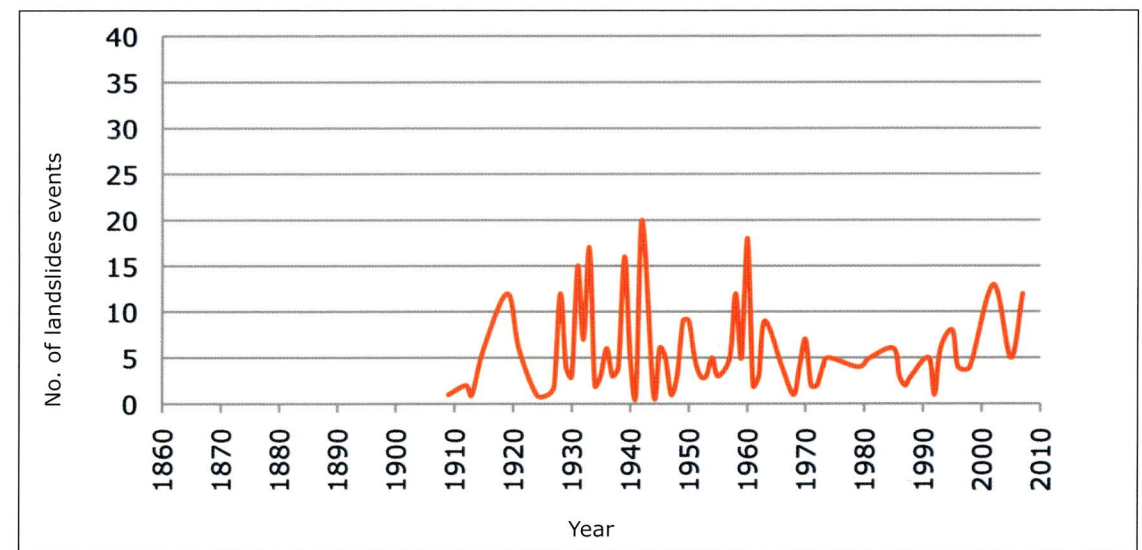

Graph 3.3.1 St Thomas landslides

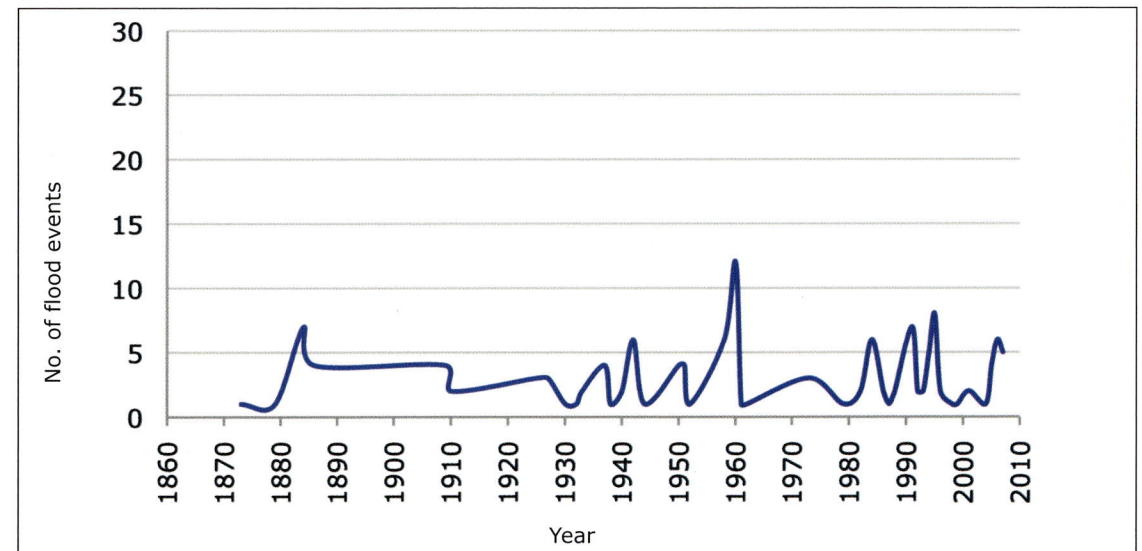

Graph 3.3.2 St Thomas floods

Plate 3.3.1 Yallahs fording flood damage due to sediment flooding (R. Ahmad, 2002)

Plate 3.3.3 Debris flow following Tropical Storm Dennis, Ramble, St Thomas (R. Ahmad, 2005)

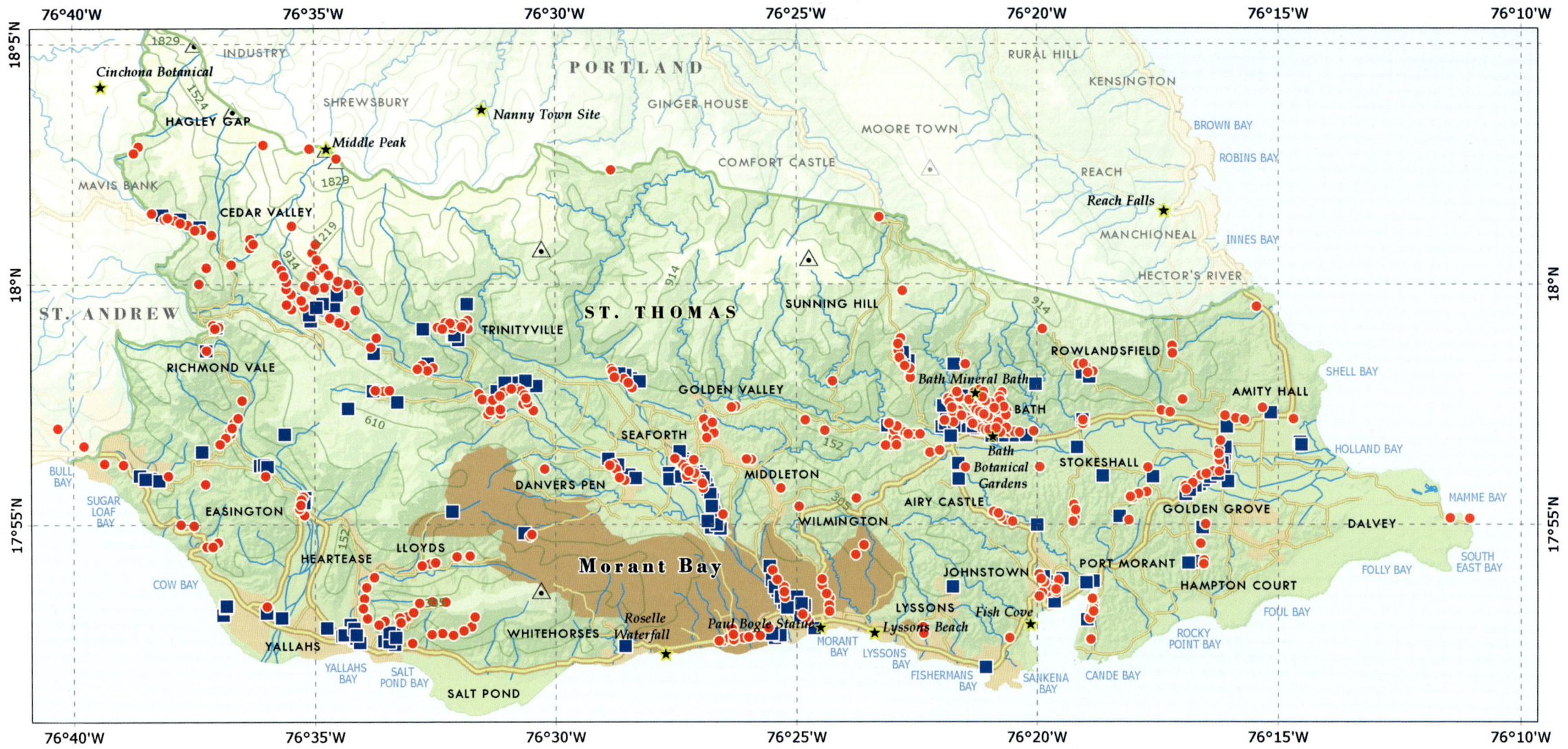

Legend:

- ● Landslide event
- ■ Flood event
- ★ Place of interest
- Parish capital
- Urban and industrial area
- Main road
- Other and parochial road
- Contours (metres)
- Major river
- △ Trigonometrical station

Elevation (metres)
- 0–500
- 501–1,000
- 1,001–2,258

1:170,000

0   2.5   5 kilometres

0   1.25   2.5   5 miles

Map Datum:
Latitude-Longitude geodetic grid
World Geodetic System 1984 datum
Prime Meridian: Greenwich
Angular unit: Degree

Map 3.3.1 St Thomas landslide and flood distribution

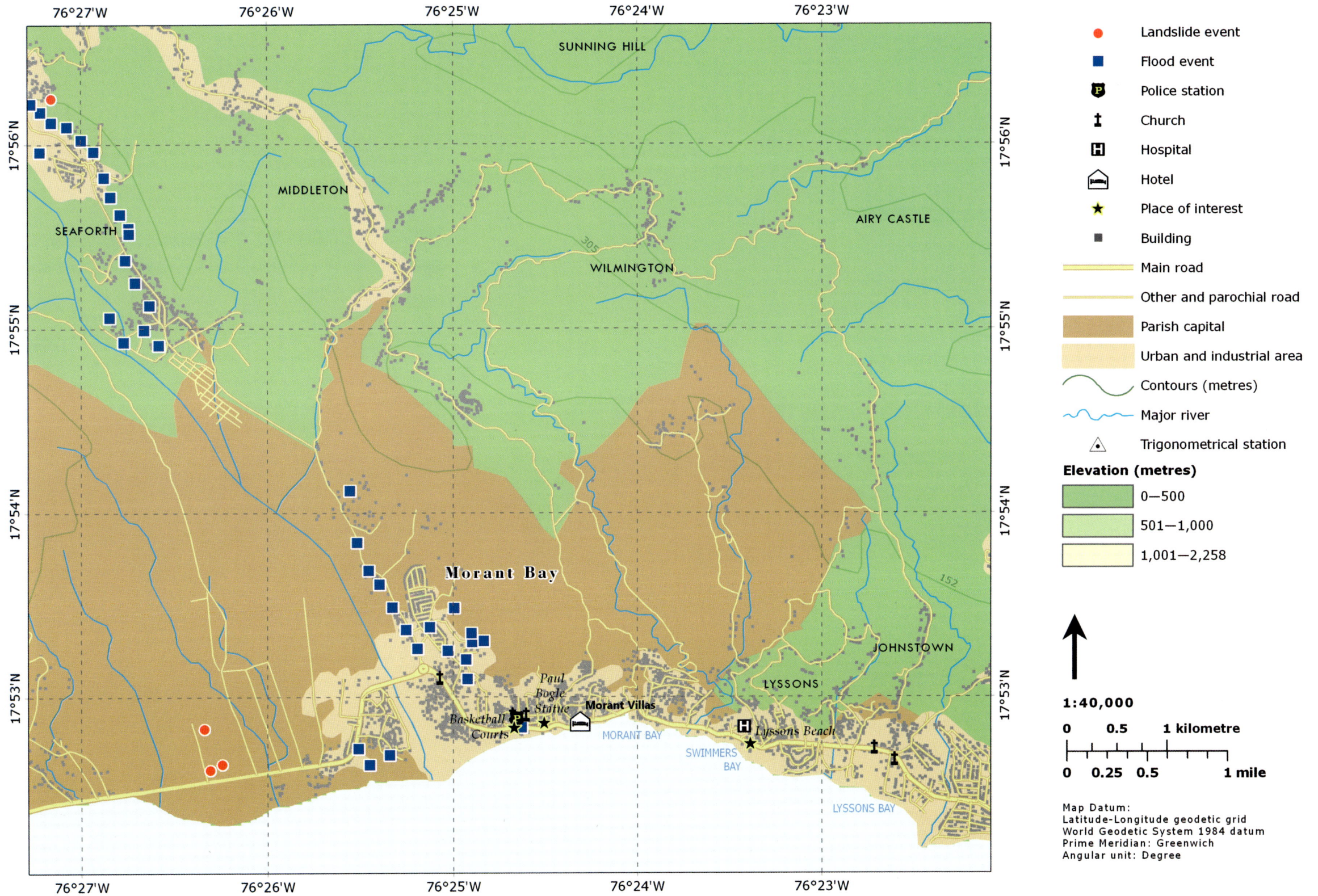

Map 3.3.2 Morant Bay landslide and flood distribution

Legend:

- ● Landslide event
- ■ Flood event
- Ⓟ Police station
- ☩ Church
- Ⓗ Hospital
- ⌂ Hotel
- ★ Place of interest
- ■ Building

Main road	
Other and parochial road	
Parish capital	
Urban and industrial area	
Contours (metres)	
Major river	
Trigonometrical station	

Elevation (metres)

- 0–500
- 501–1,000
- 1,001–2,258

1:40,000

0   0.5   1 kilometre

0   0.25   0.5   1 mile

Map Datum:
Latitude-Longitude geodetic grid
World Geodetic System 1984 datum
Prime Meridian: Greenwich
Angular unit: Degree

3.4 PORTLAND

PARISH PROFILE

Human geography: The population of Portland is around 80,000 people. Agriculture and tourism dominate, with coffee production along the slopes of the Blue Mountains, and tourism along the coast. The capital of Portland is Port Antonio.

Physical geography: Northern slopes of the highly dissected Blue Mountains are prone to landslides. Orographic rainfall is influenced by the presence of the Blue Mountains to the south of the parish. This leads to a great deal of surface run-off, forming significant rivers in the parish. Alluvial fans are found at the mouths of all the major rivers.

Hazard profile: Flooding potential is very high at the mouths of the rivers, while landslides are common to the south along the slopes of the Blue Mountains. Road breakaways associated with landslides commonly cut off entire communities.

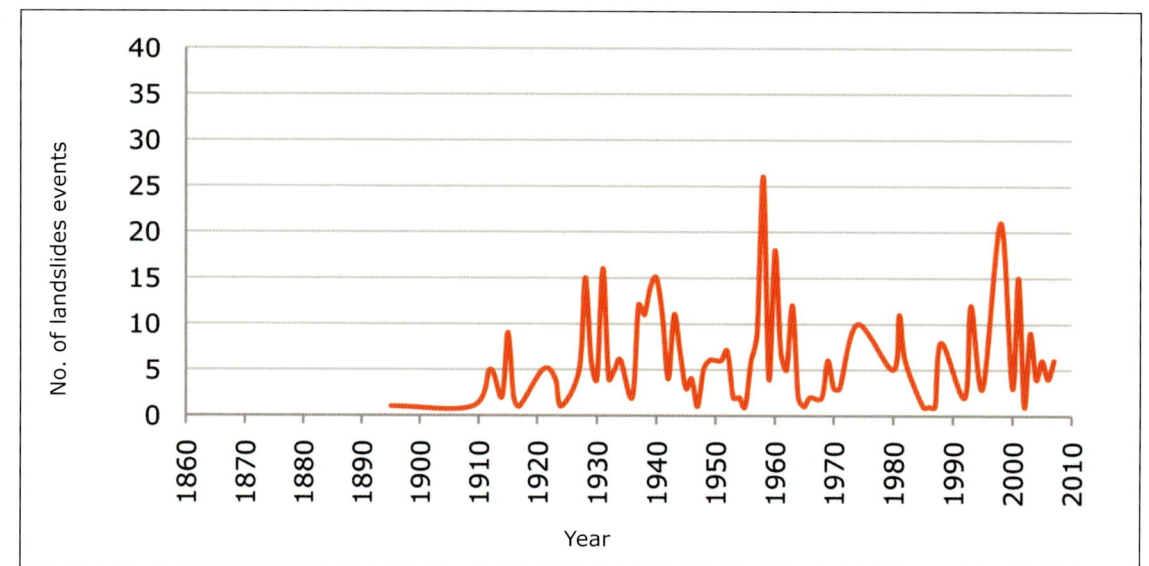

Graph 3.4.1 Portland landslides

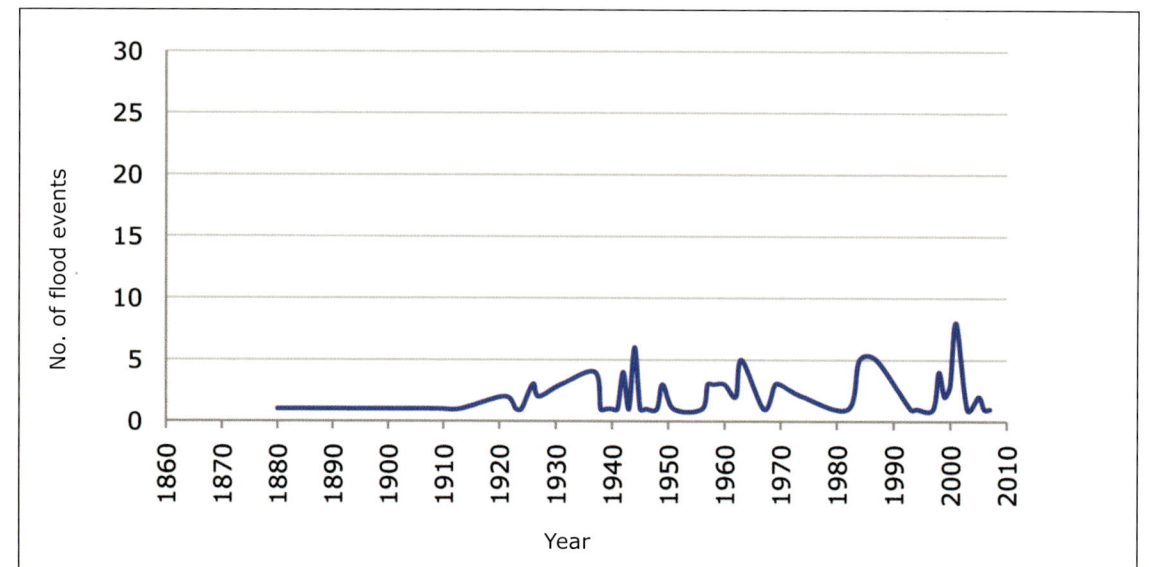

Graph 3.4.2 Portland floods

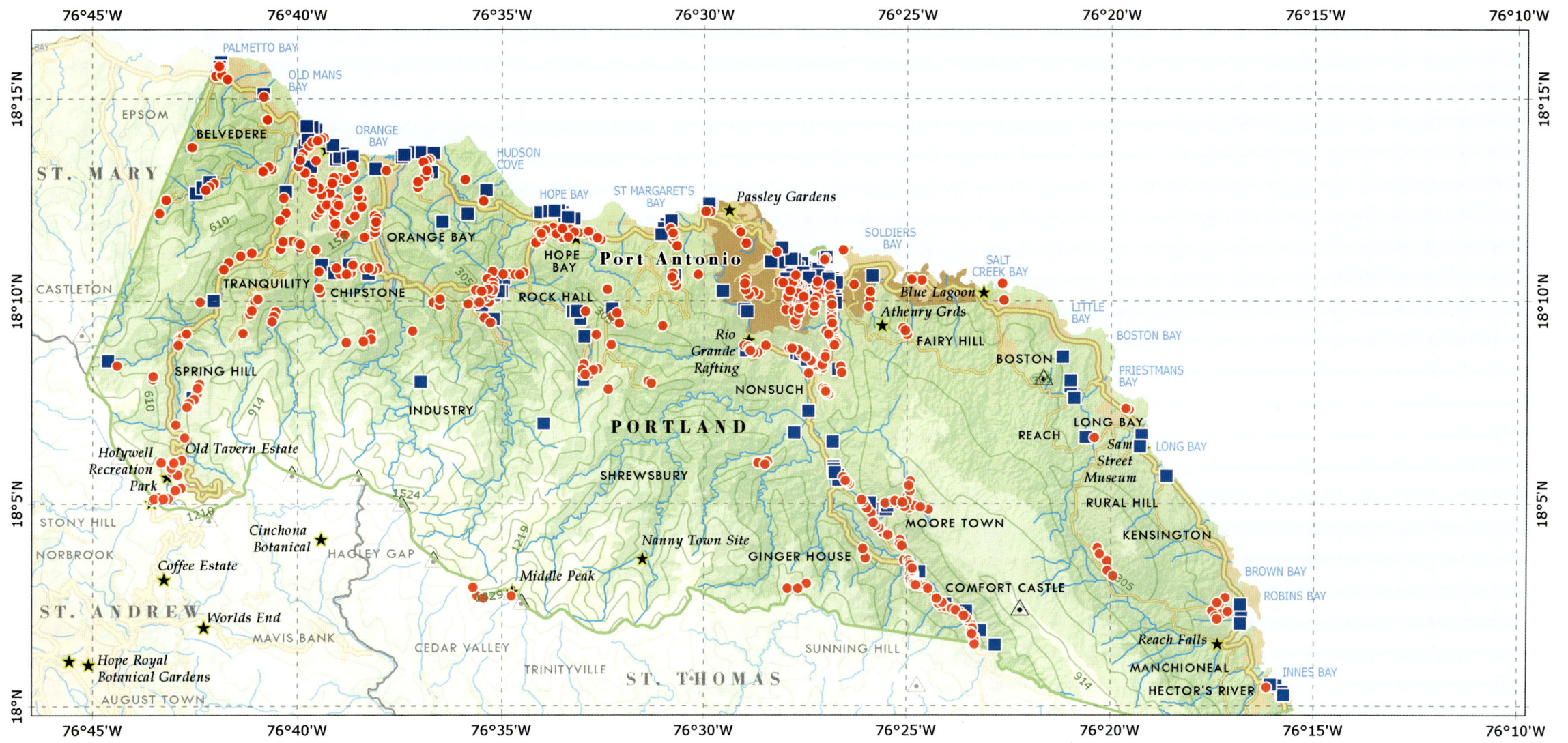

Map 3.4.1 Portland landslide and flood distribution

Legend

- ● Landslide event
- ■ Flood event
- ★ Place of interest
- Parish capital
- Urban and industrial area

- Main road
- Other and parochial road
- Contours (metres)
- Major river
- ⬛ Trigonometrical station

Elevation (metres)
- 0—500
- 501—1,000
- 1,001—2,258

1:200,000

0   2.5   5 kilometres

0   1.25   2.5   5 miles

Map Datum:
Latitude-Longitude geodetic grid
World Geodetic System 1984 datum
Prime Meridian: Greenwich
Angular unit: Degree

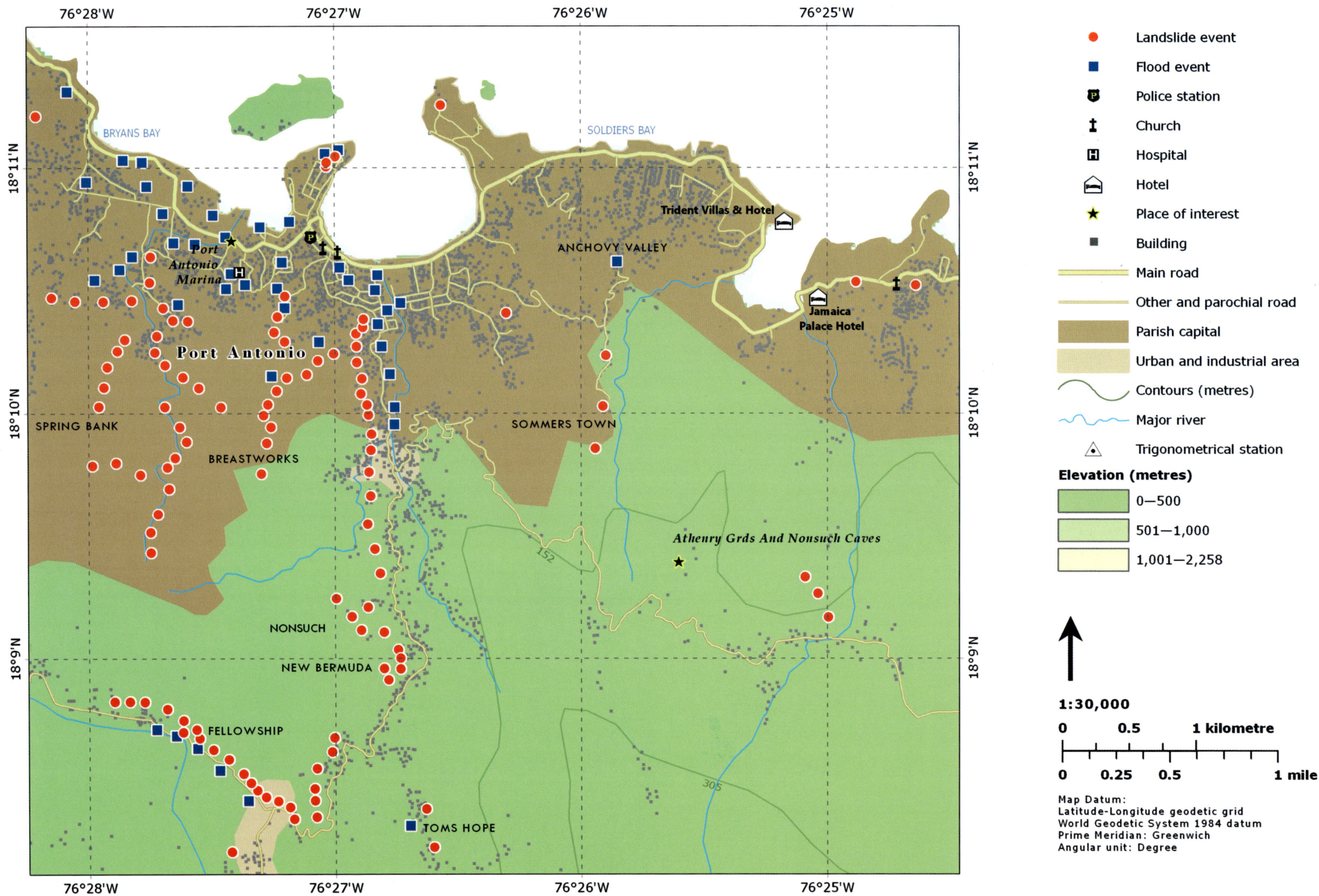

Map 3.4.2 Port Antonio landslide and flood distribution

Legend:
- ● Landslide event
- ■ Flood event
- Police station
- Church
- H Hospital
- Hotel
- ★ Place of interest
- ■ Building
- Main road
- Other and parochial road
- Parish capital
- Urban and industrial area
- Contours (metres)
- Major river
- △ Trigonometrical station

Elevation (metres)
- 0—500
- 501—1,000
- 1,001—2,258

1:30,000

0   0.5   1 kilometre

0   0.25   0.5   1 mile

Map Datum:
Latitude-Longitude geodetic grid
World Geodetic System 1984 datum
Prime Meridian: Greenwich
Angular unit: Degree

Plate 3.4.1 Debris flows following 2001 rainfall blocked and overtopped the bridge near Belcarres, Buff Bay Valley, Portland (R. Ahmad, 2001)

Plate 3.4.2 Bybrook, flood rains, October 2001 (R. Ahmad, 2001)

Plate 3.4.3 Belcarres debris flow, October 2001 (R. Ahmad, 2001)

3.5   ST MARY

PARISH PROFILE

Human geography: The population of St Mary is over 110,000 people. Economic activity is largely agriculture, particularly banana production. The capital of St Mary is Port Maria.

Physical geography: The parish consists largely of clay bedrock; uplands are dominated by limestone, particularly in the west. There is a high drainage density in the parish.

Hazard profile: St Mary is highly susceptible to landslides and flooding. The parish has the highest number of reported landslides in the *Gleaner* archives, owing largely to the frequency of landslides along the main road from Annotto Bay to Kingston, where landslides and debris flows are very common.

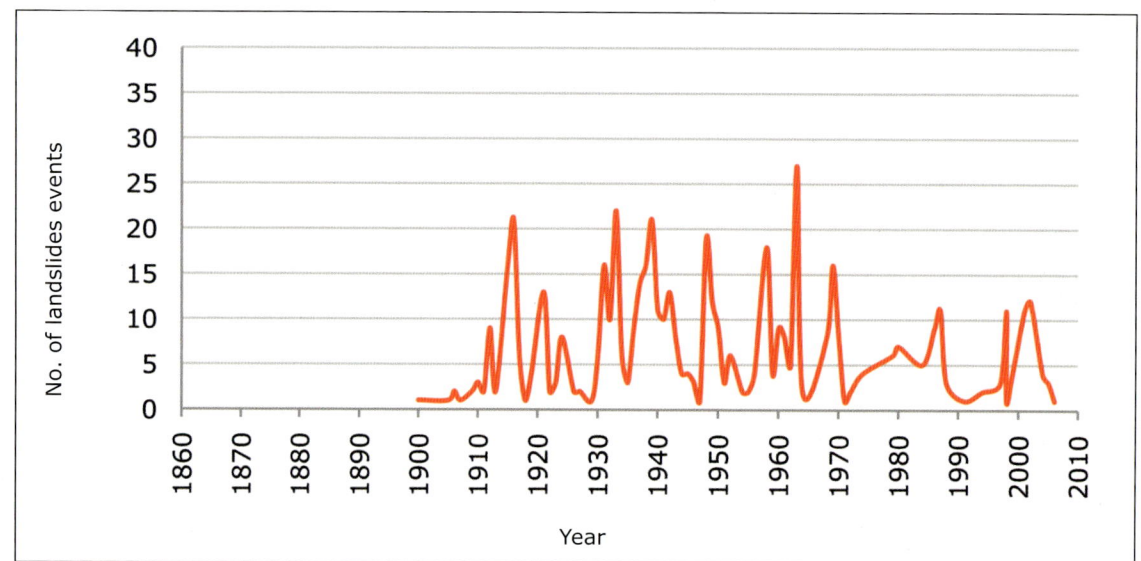

Graph 3.5.1 St Mary landslides

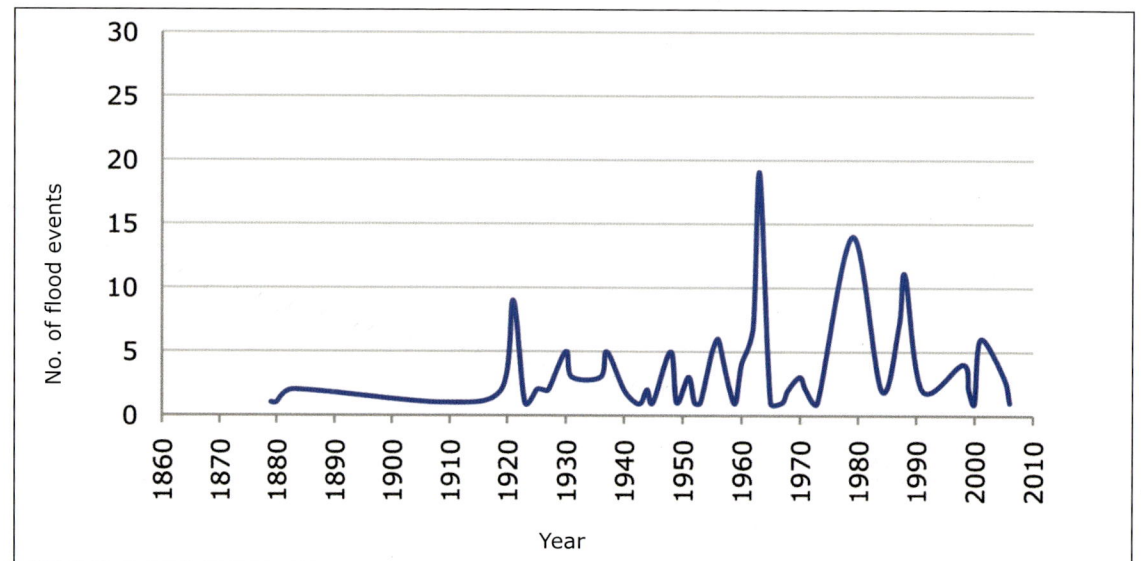

Graph 3.5.2 St Mary floods

Plate 3.5.1 Whitehall land slippage, 2005 (P. Lyew-Ayee Jr, 2005)

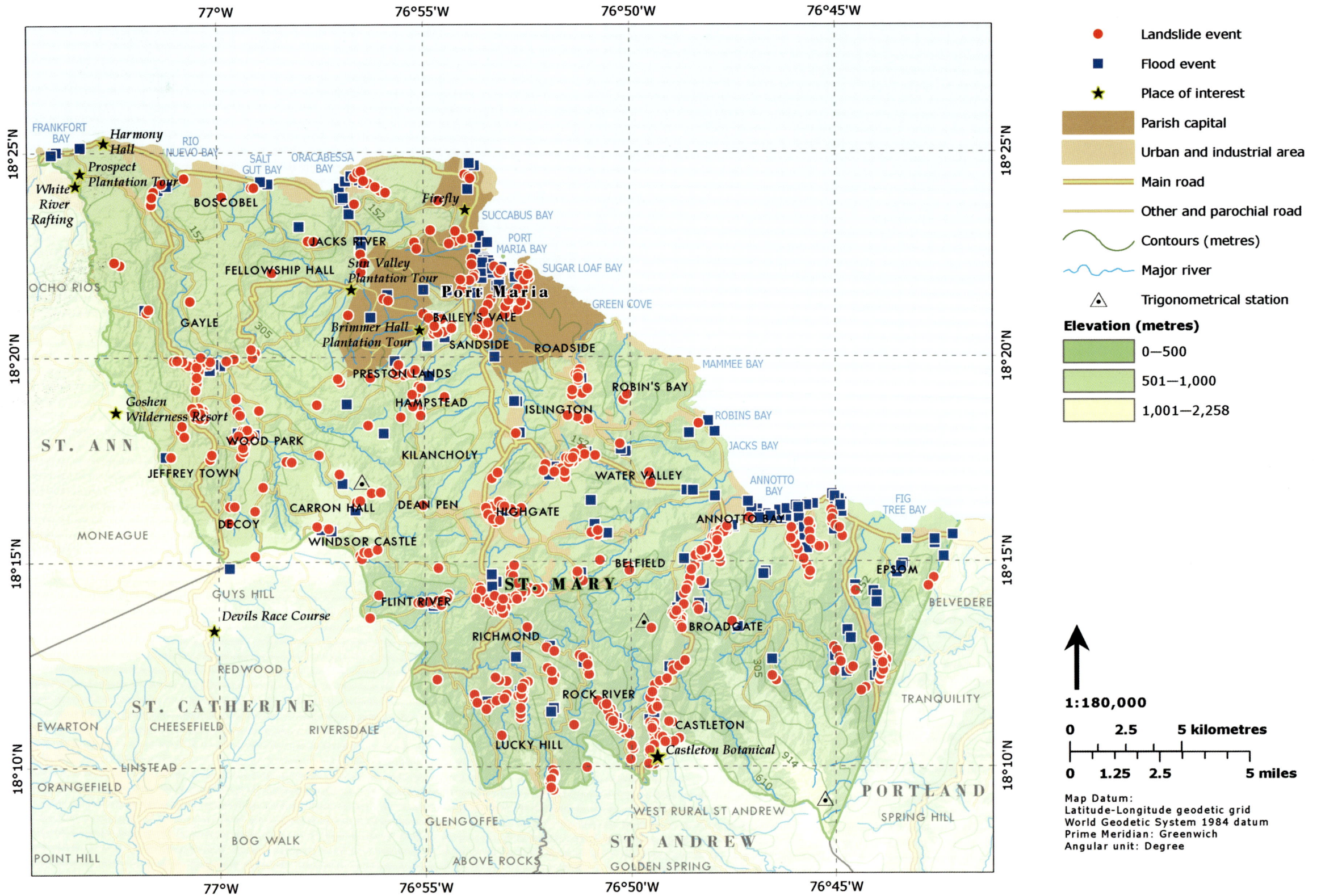

Map 3.5.1 St Mary landslide and flood distribution

Legend

- Landslide event
- Flood event
- Place of interest
- Parish capital
- Urban and industrial area
- Main road
- Other and parochial road
- Contours (metres)
- Major river
- Trigonometrical station

Elevation (metres)

- 0—500
- 501—1,000
- 1,001—2,258

1:180,000

0   2.5   5 kilometres

0   1.25   2.5   5 miles

Map Datum:
Latitude-Longitude geodetic grid
World Geodetic System 1984 datum
Prime Meridian: Greenwich
Angular unit: Degree

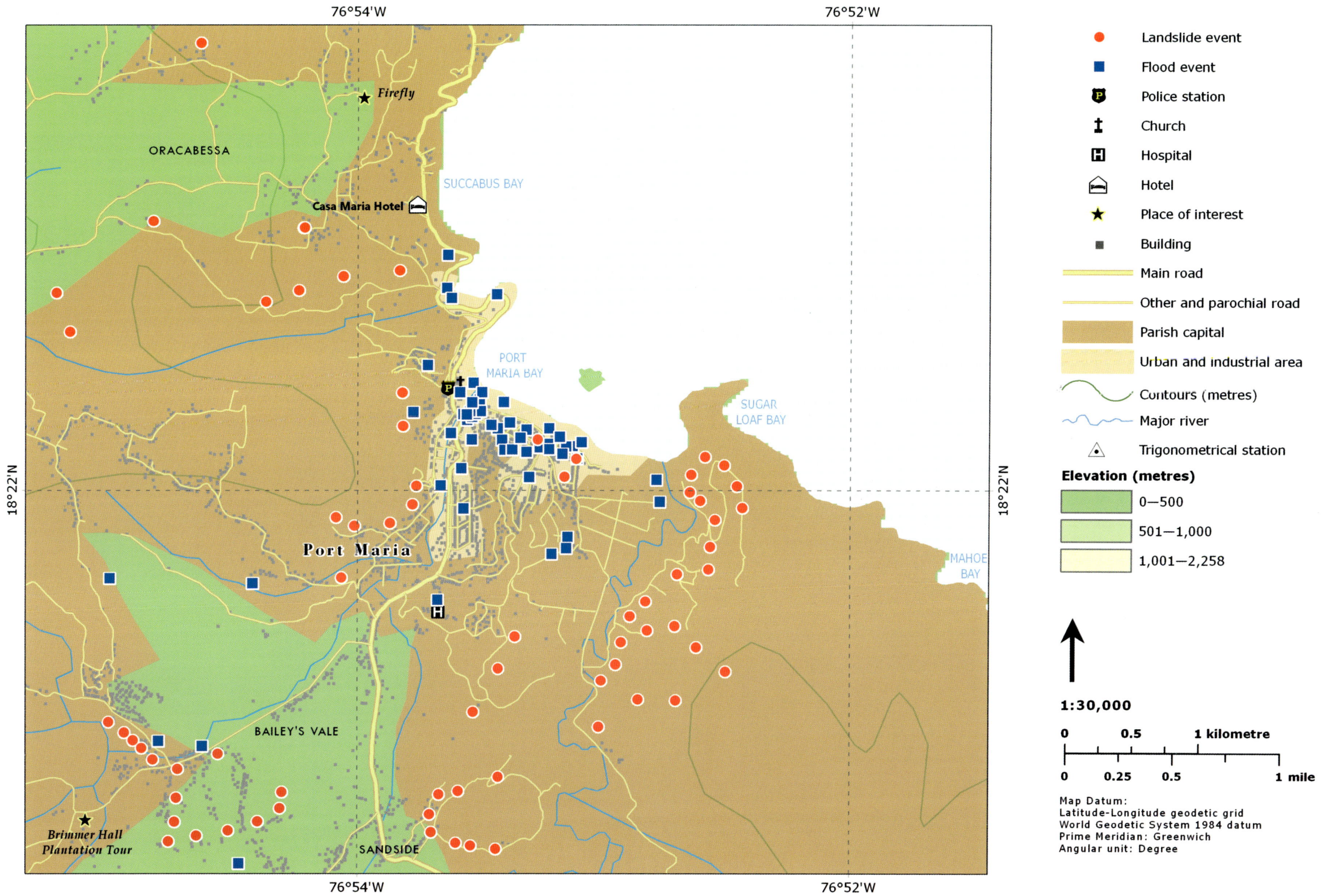

Legend

- 🔴 Landslide event
- 🟦 Flood event
- Ⓟ Police station
- ✝ Church
- Ⓗ Hospital
- ⌂ Hotel
- ★ Place of interest
- ▪ Building
- Main road
- Other and parochial road
- Parish capital
- Urban and industrial area
- Contours (metres)
- Major river
- △ Trigonometrical station

Elevation (metres)

- 0—500
- 501—1,000
- 1,001—2,258

1:30,000

| 0 | 0.5 | 1 kilometre |

| 0 | 0.25 | 0.5 | 1 mile |

Map Datum:
Latitude-Longitude geodetic grid
World Geodetic System 1984 datum
Prime Meridian: Greenwich
Angular unit: Degree

Map 3.5.2 Port Maria landslide and flood distribution

3.6 ST ANN

PARISH PROFILE

Human geography: The population of St Ann is over 160,000 people. Economic activities include agriculture (cattle farming and, inland, small farming), tourism (along the coast, but particularly in the Ocho Rios, Runaway Bay and Discovery Bay areas) and bauxite mining (largely in central St Ann, with a plant along the coast). The capital of St Ann is St Ann's Bay.

Physical geography: The parish has largely karst landscape, with the Dry Harbour Mountains to the south, with its degraded karst terrain. The large E–W Duanvale fault system crosses the parish. To the north, doline karst landscapes and raised reefs are found. There are few large river systems owing to the karst landscape, but many springs.

Hazard profile: Frequent karst flooding occurs in areas close to the water table, particularly in Moneague in the east and Cave Valley to the south. Localized karst flooding also occurs in periods of intense and prolonged rainfall. Landslides are not as common in limestone areas, although rock falls may occur in structurally weak areas. The town of Ocho Rios is susceptible to debris floods due to its location along a debris fan, exacerbated by the fact that the main channel, Fern Gully, has been paved over into a major thoroughfare, and illegal settlements in that sub-watershed have led to deforestation and, consequently, increased surface run-off.

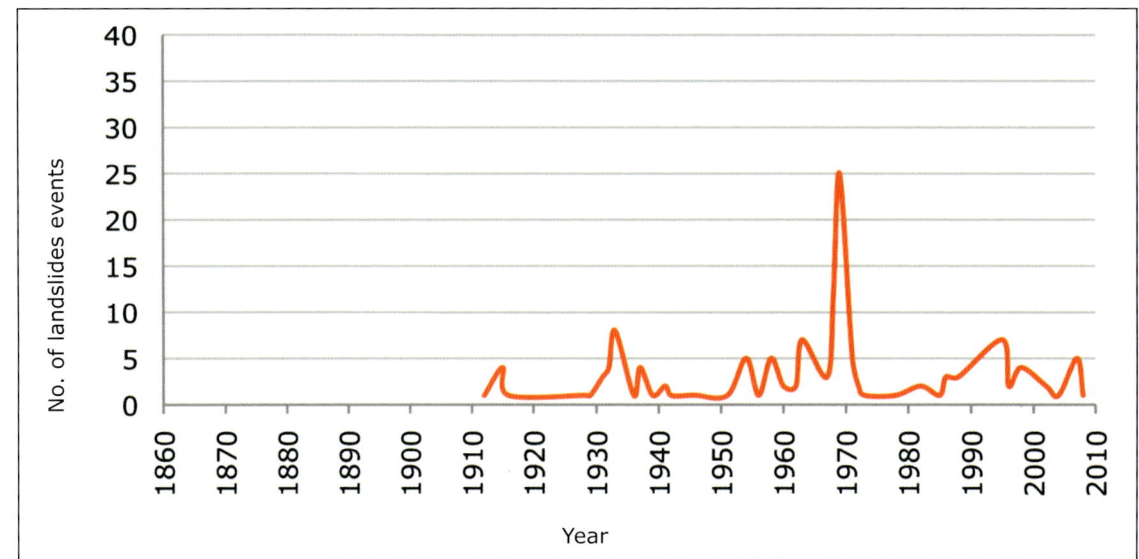

Graph 3.6.1 St Ann landslides

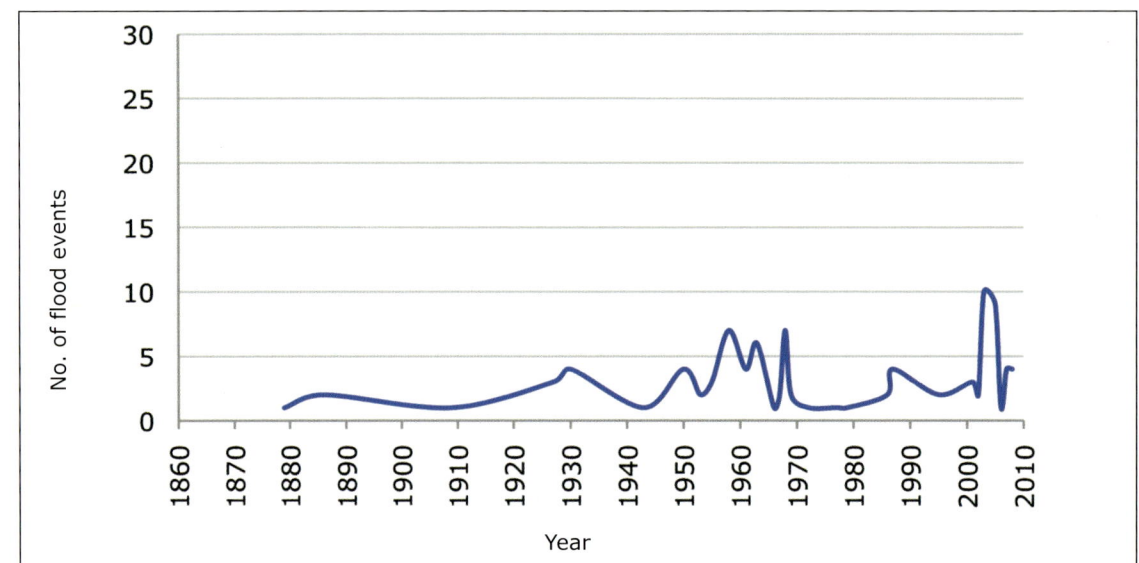

Graph 3.6.2 St Ann floods

Plate 3.6.1 Rio Hoe, St Ann, flood rains, April 2008 (Water Resources Authority, 2008)

DUNCANS

RIO BUENO

SAMUELS PROSPECT

Columbus Park ★

RUNAWAY BAY

SPICY HILL

Green Grotto Caves

DISCOVERY BAY

Cardiff Hall

Chukka Cove

ST ANNS BAY

JACKSON TOWN

SAWYERS

RUNAWAY BAY

BEVERLY

MAMMEE BAY

18°25'N

LAWRENCE PARK

Circle B Farm

Seville Great House ★

OCHO RIOS BAY

WHITE RIVER BAY

Harmony Hall ★

RIO NUEVO BAY

SALT GUT BAY

CLARKS TOWN

STURGE TOWN

St Anns Bay

Dunns River Falls

Prospect Plantation Tours ★
White River Rafting ★

SHERWOOD CONTENT

Roaring River Falls

STEER TOWN

EXCHANGE

ELTHAM

BOSCOBEL

BAMBOO

OCHO RIOS △

Fern Gully

JACKS RIVER

TRELAWNY

BROWNS TOWN

ST. MARY

FELLOWSHIP HALL

ROCK SPRING

Ramgoat Cave ★

305

914

ST. ANN

18°20'N

GOLDEN GROVE

GAYLE

ULSTER SPRING

WATT TOWN

ST DACRE

WALKERS WOOD

Goshen Wilderness Resort ★

WOOD PARK

ALBERT TOWN

FREEMAN'S HALL

ALEXANDRA

Bob Marley Mausoleum

HELLSHIRE

610

CARRON HALL

WARSOP

CLAREMONT

DECOY

18°15'N

LITCHFIELD

WAIT A BIT

610

MONEAGUE

GUYS HILL

LOWE RIVER

PILE

914

CAVE VALLEY

BENSONTON

610

Devils Race Course ★

Gourie State Park ★

CASCADE

ST. CATHERINE

MANCHESTER

AENON TOWN

CLARENDON   JAMES HILL

LLUIDAS VALE

EWARTON

LINSTEAD

CHEESEFIELD

Legend

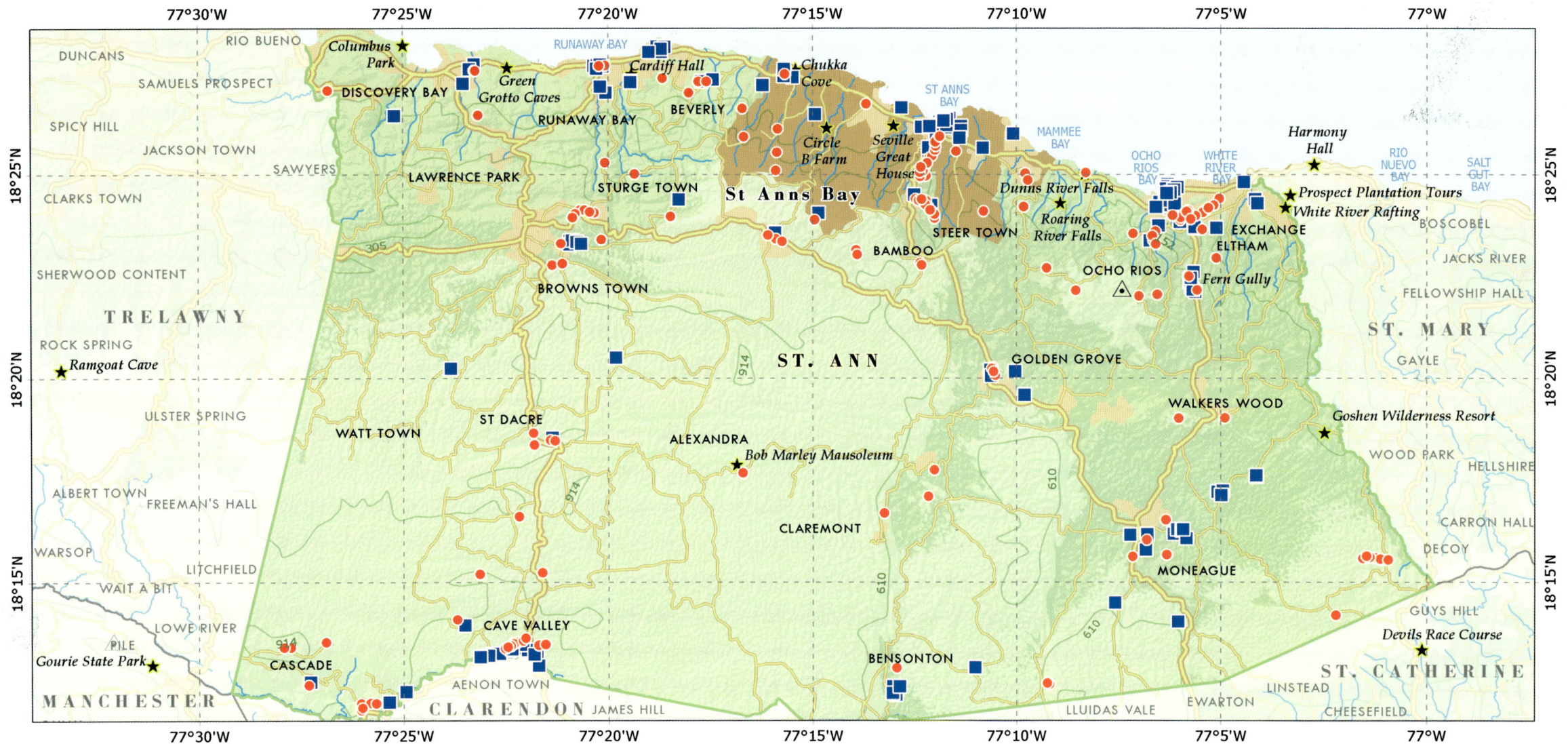

- ● Landslide event
- ■ Flood event
- ★ Place of interest
- Parish capital
- Urban and industrial area
- Main road
- Other and parochial road
- Contours (metres)
- Major river
- △ Trigonometrical station

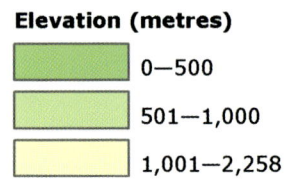

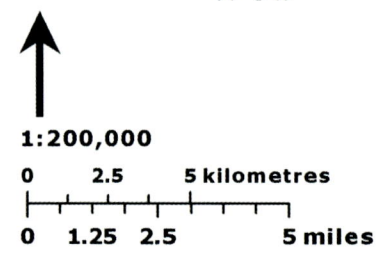

Elevation (metres)

- 0—500
- 501—1,000
- 1,001—2,258

1:200,000

0   2.5   5 kilometres

0   1.25   2.5   5 miles

Map Datum:
Latitude-Longitude geodetic grid
World Geodetic System 1984 datum
Prime Meridian: Greenwich
Angular unit: Degree

Map 3.6.1 St Ann landslide and flood distribution

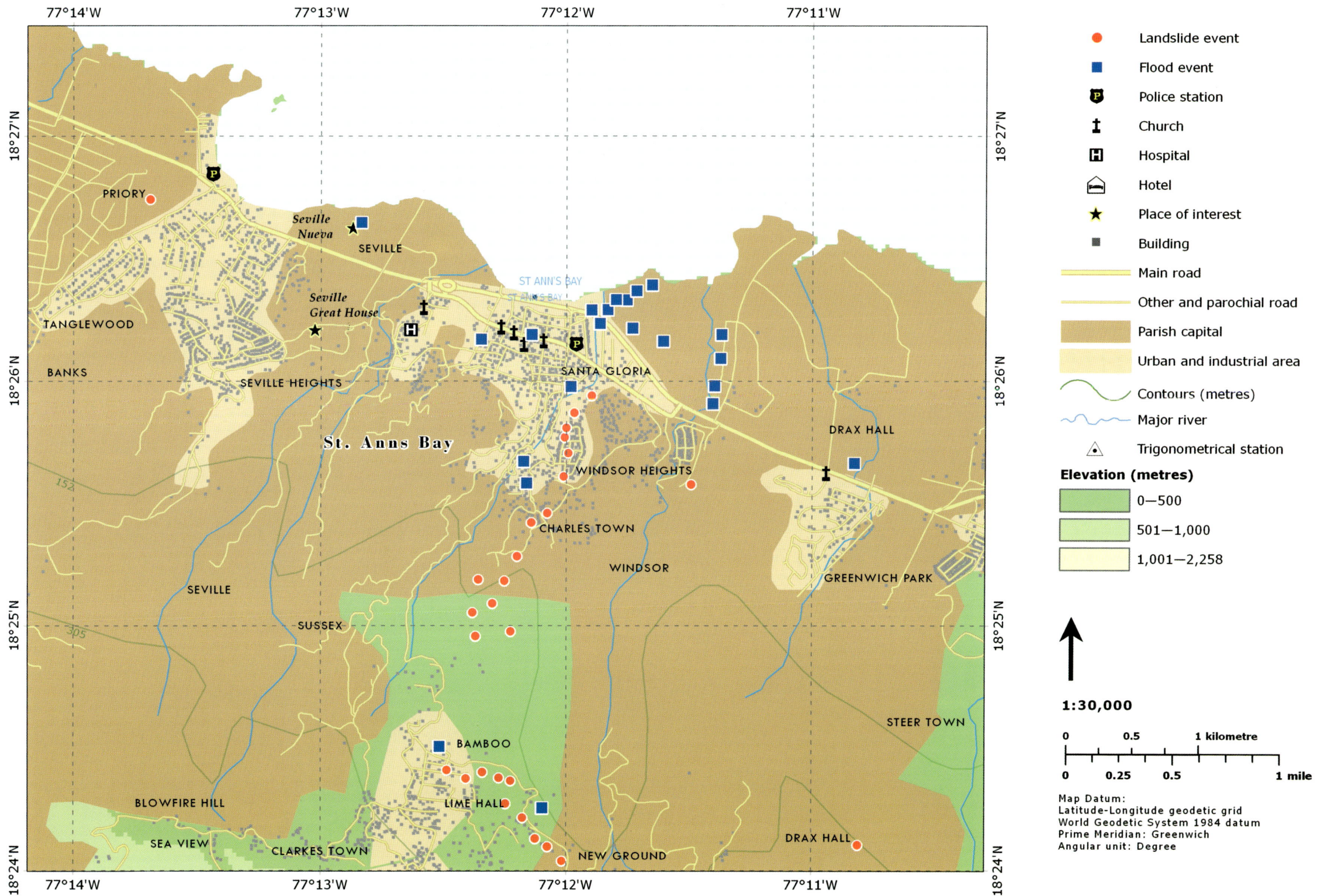

Map 3.6.2 St Ann's Bay landslide and flood distribution

Legend

- 🔴 Landslide event
- 🟦 Flood event
- Ⓟ Police station
- ✝ Church
- Ⓗ Hospital
- 🛏 Hotel
- ★ Place of interest
- ▪ Building
- ▬ Main road
- ▬ Other and parochial road
- Parish capital
- Urban and industrial area
- Contours (metres)
- Major river
- △ Trigonometrical station

Elevation (metres)

- 0—500
- 501—1,000
- 1,001—2,258

1:30,000

0   0.5   1 kilometre

0   0.25   0.5   1 mile

Map Datum:
Latitude-Longitude geodetic grid
World Geodetic System 1984 datum
Prime Meridian: Greenwich
Angular unit: Degree

Landslide event

Flood event

Police station

Church

Hospital

Hotel

Place of interest

Building

Main road

Other and parochial road

Parish capital

Urban and industrial area

Contours (metres)

Major river

Trigonometrical station

Elevation (metres)

0—500

501—1,000

1,001—2,258

1:25,000

0   0.5   1 kilometre

0   0.25   0.5   1 mile

Map Datum:
Latitude-Longitude geodetic grid
World Geodetic System 1984 datum
Prime Meridian: Greenwich
Angular unit: Degree

77°7'W   77°6'W   77°5'W   77°4'W

18°26'N

18°25'N

18°24'N

MALLARDS BAY

Silver Seas Hotel

Sunset Jamaica Grande Hotel

Hibiscus Lodge Hotel

OCHO RIOS BAY

Fisherman's Point

Turtle Beach Towers

Craine Ridge Resort

SHAW PARK

The Enchanted Golden Resort

Ochio Rios

Simanda Villa

Pineapple Hotel

Sandals

WHITE RIVER BAY

GAYLE

Couples, San Souci

Shaw Park Beach Hotel

Rio Blanco Apartments

SANDY BEACH BAY

Royal Plantation Golf Resort & Spa

Sandals Grande Manor

CONTENT GARDEN

EXCHANGE

ELTHAM

STANMORE GROVE

HARRISON TOWN

GREAT POND

PARRY TOWN

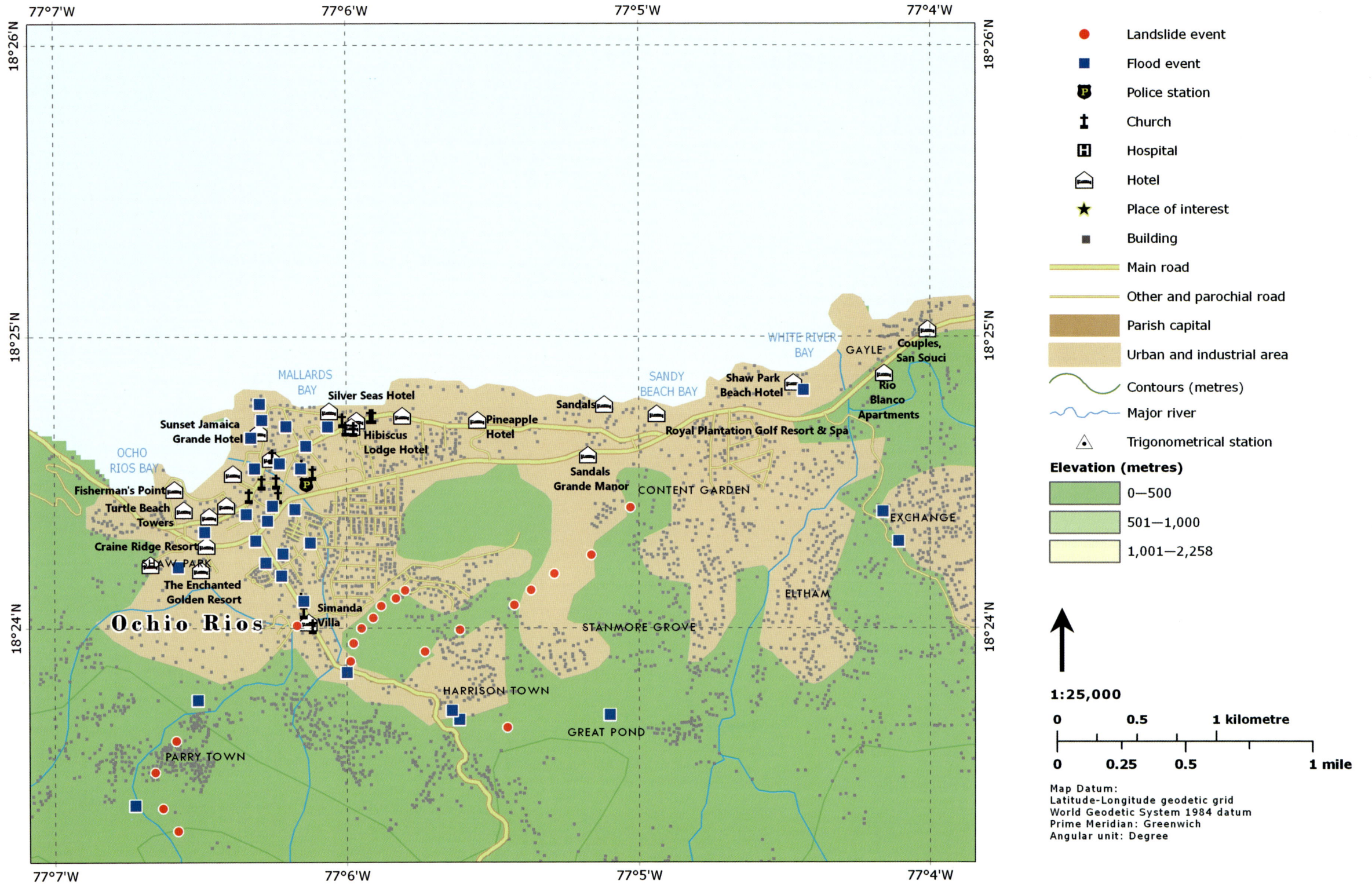

Map 3.6.3 Ocho Rios landslide and flood distribution

Plate 3.6.2 Main Street, Ocho Rios, flood rains, April 2008 (Water Resources Authority, 2008)

3.7 TRELAWNY

PARISH PROFILE

Human geography: The population of Trelawny is just under 75,000. Main economic activities include agriculture (sugar-cane plantation at Long Pond, as well as small farming inland) and tourism (along the coast). The capital of Trelawny is Falmouth.

Physical geography: The parish has largely karst landscape, with the Cockpit Country to the south, with classic cockpit karst terrain. To the north of the Duanvale fault system, coastal karst landscape is found. The Martha Brae river system ends at a swampy delta near Falmouth.

Hazard profile: The parish experiences frequent karst flooding along areas near the water table, especially along the northern boundary of the Cockpit Country. Flooding is also likely along the low-lying swampy delta of the Martha Brae river. Landslides are not common, although rock falls may occur in structurally weak areas.

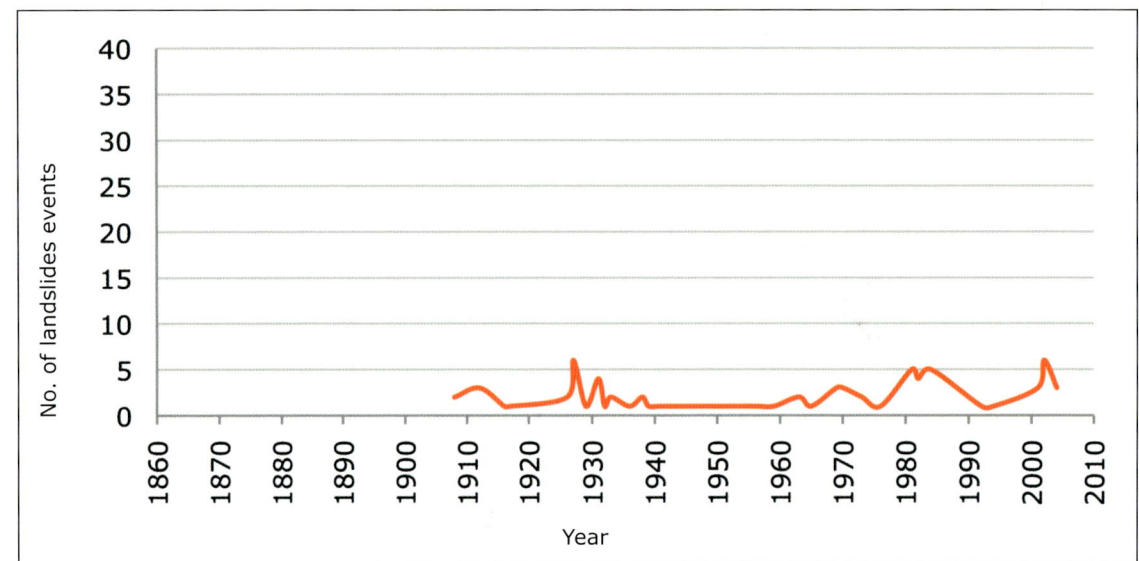

Graph 3.7.1 Trelawny landslides

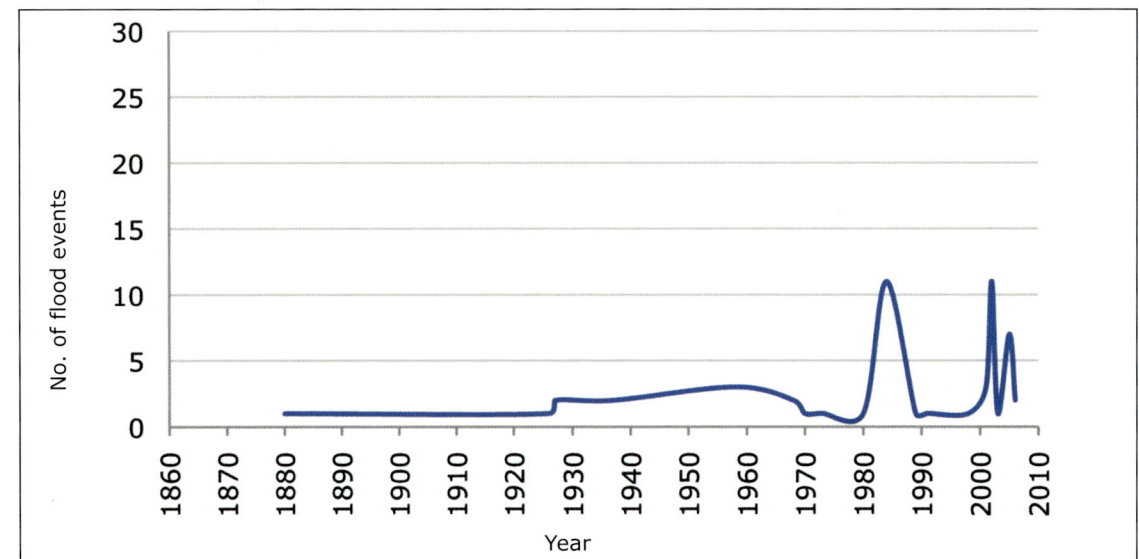

Graph 3.7.2 Trelawny floods

Plate 3.7.1 Cornwall Street flooding, Falmouth, Trelawny (*Gleaner*, 2010)

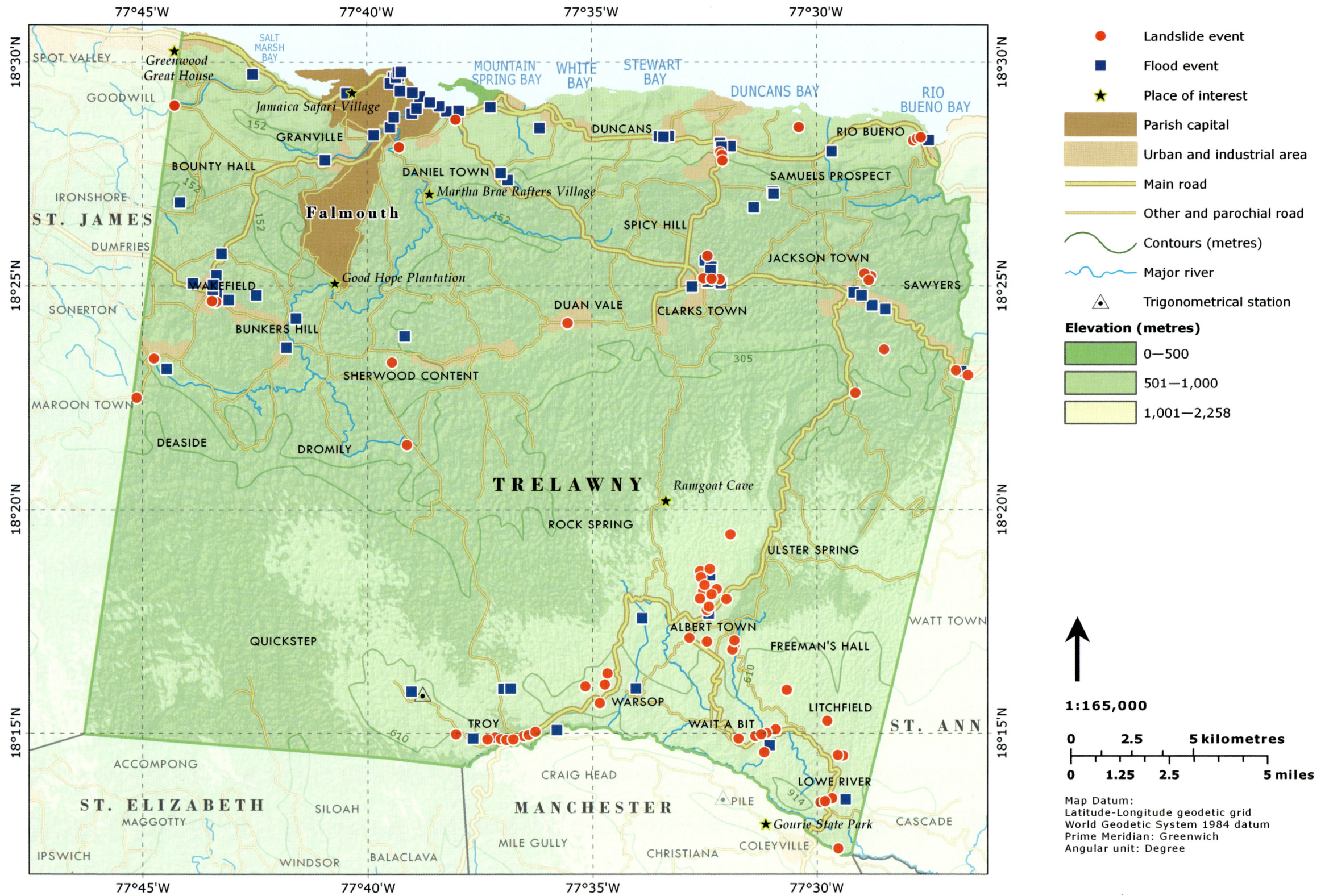

Map 3.7.1 Trelawny landslide and flood distribution

Legend:
- Landslide event
- Flood event
- Place of interest
- Parish capital
- Urban and industrial area
- Main road
- Other and parochial road
- Contours (metres)
- Major river
- Trigonometrical station

Elevation (metres)
- 0—500
- 501—1,000
- 1,001—2,258

1:165,000

0   2.5   5 kilometres

0   1.25   2.5   5 miles

Map Datum:
Latitude-Longitude geodetic grid
World Geodetic System 1984 datum
Prime Meridian: Greenwich
Angular unit: Degree

Map labels:
SPOT VALLEY, Greenwood Great House, SALT MARSH BAY, MOUNTAIN SPRING BAY, WHITE BAY, STEWART BAY, DUNCANS BAY, RIO BUENO BAY, GOODWILL, Jamaica Safari Village, GRANVILLE, DUNCANS, RIO BUENO, BOUNTY HALL, DANIEL TOWN, SAMUELS PROSPECT, IRONSHORE, ST. JAMES, Martha Brae Rafters Village, DUMFRIES, Falmouth, SPICY HILL, JACKSON TOWN, WAKEFIELD, Good Hope Plantation, DUAN VALE, CLARKS TOWN, SAWYERS, SONERTON, BUNKERS HILL, MAROON TOWN, SHERWOOD CONTENT, DEASIDE, DROMILY, TRELAWNY, Ramgoat Cave, ROCK SPRING, ULSTER SPRING, WATT TOWN, QUICKSTEP, ALBERT TOWN, FREEMAN'S HALL, WARSOP, LITCHFIELD, ST. ANN, TROY, WAIT A BIT, ACCOMPONG, CRAIG HEAD, PILE, LOWE RIVER, ST. ELIZABETH, MANCHESTER, MAGGOTTY, SILOAH, MILE GULLY, CHRISTIANA, COLEYVILLE, CASCADE, Gourie State Park, IPSWICH, WINDSOR, BALACLAVA

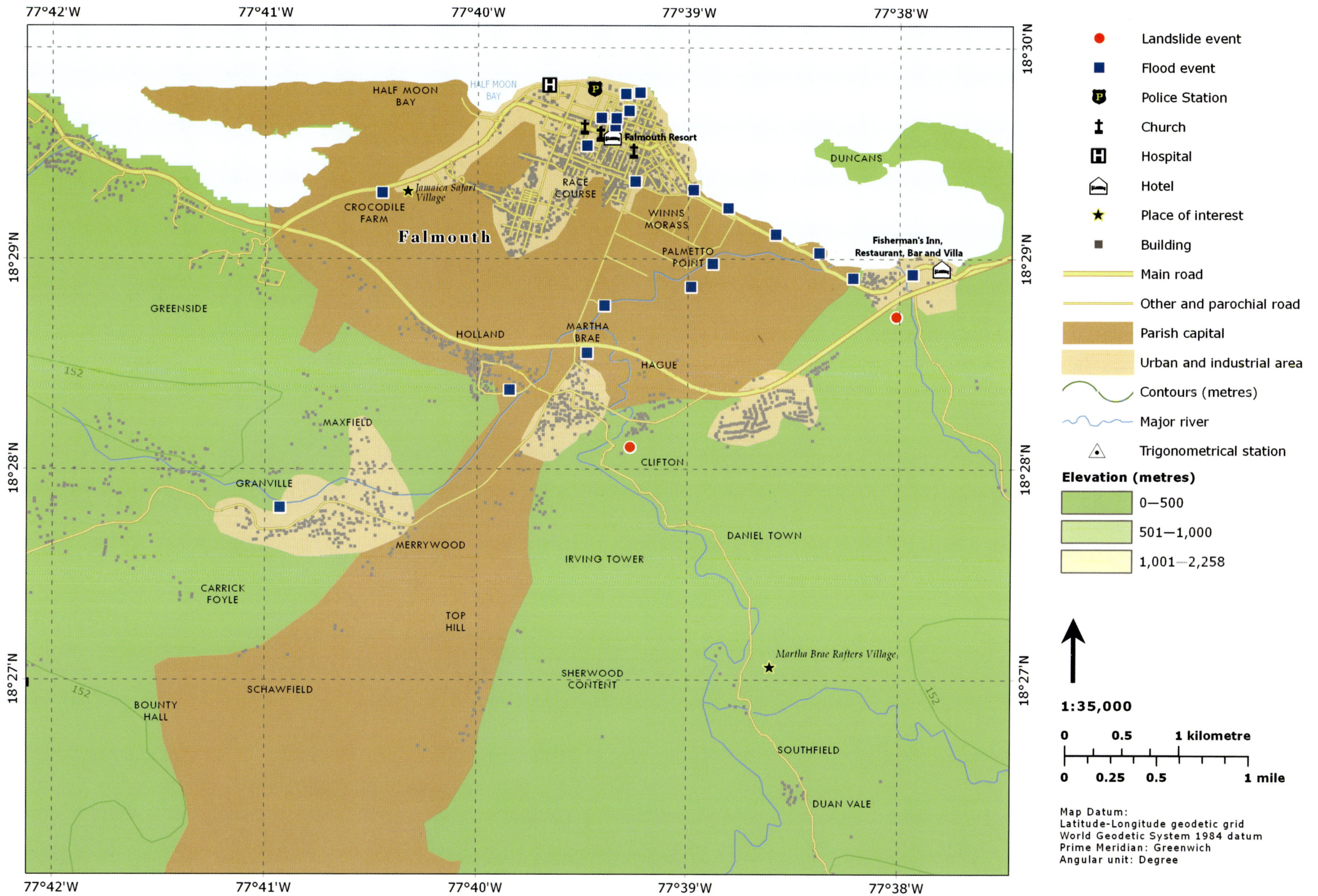

Map 3.7.2 Falmouth landslide and flood distribution

Legend

- ● Landslide event
- ■ Flood event
- Ⓟ Police Station
- ✝ Church
- Ⓗ Hospital
- 🛏 Hotel
- ★ Place of interest
- ▪ Building
- Main road
- Other and parochial road
- Parish capital
- Urban and industrial area
- Contours (metres)
- Major river
- △ Trigonometrical station

Elevation (metres)

- 0—500
- 501—1,000
- 1,001—2,258

1:35,000

| 0 | 0.5 | 1 kilometre |

| 0 | 0.25 | 0.5 | 1 mile |

Map Datum:
Latitude-Longitude geodetic grid
World Geodetic System 1984 datum
Prime Meridian: Greenwich
Angular unit: Degree

3.8   ST JAMES

PARISH PROFILE

Human geography: The population of St James is around 175,000. Main economic activities include manufacturing, agriculture and tourism. The capital of St James is Jamaica's second city, Montego Bay.

Physical geography: The parish has varied landscape, with a large karst polje to the southeast, near Wakefield. The Maldon and Sunderland areas are dominated by outcrops of deeply weathered volcaniclastic rocks.

Hazard profile: Frequent flooding along the karst plains of the Queen of Spain's Valley, which includes the towns of Hampden and Wakefield. Lower-lying areas in the vicinity of Montego Bay are also vulnerable to coastal flooding, and reclaimed land would also be vulnerable to earthquakes.

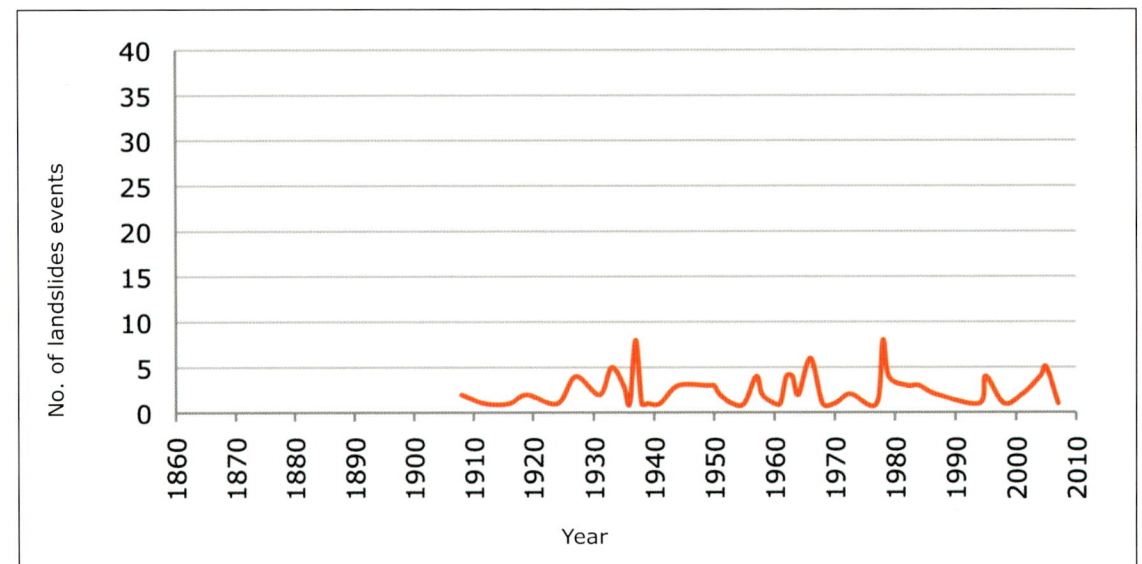

Graph 3.8.1 St James landslides

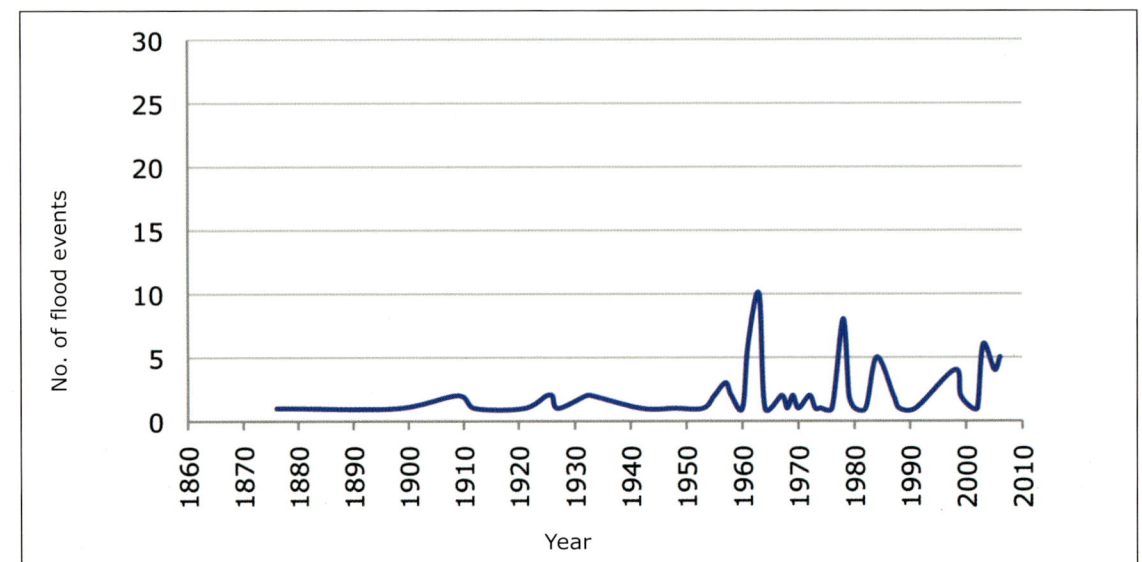

Graph 3.8.2 St James floods

Plate 3.8.1 Road destroyed by landslides, Tangle River, St James (NWA, 2004)

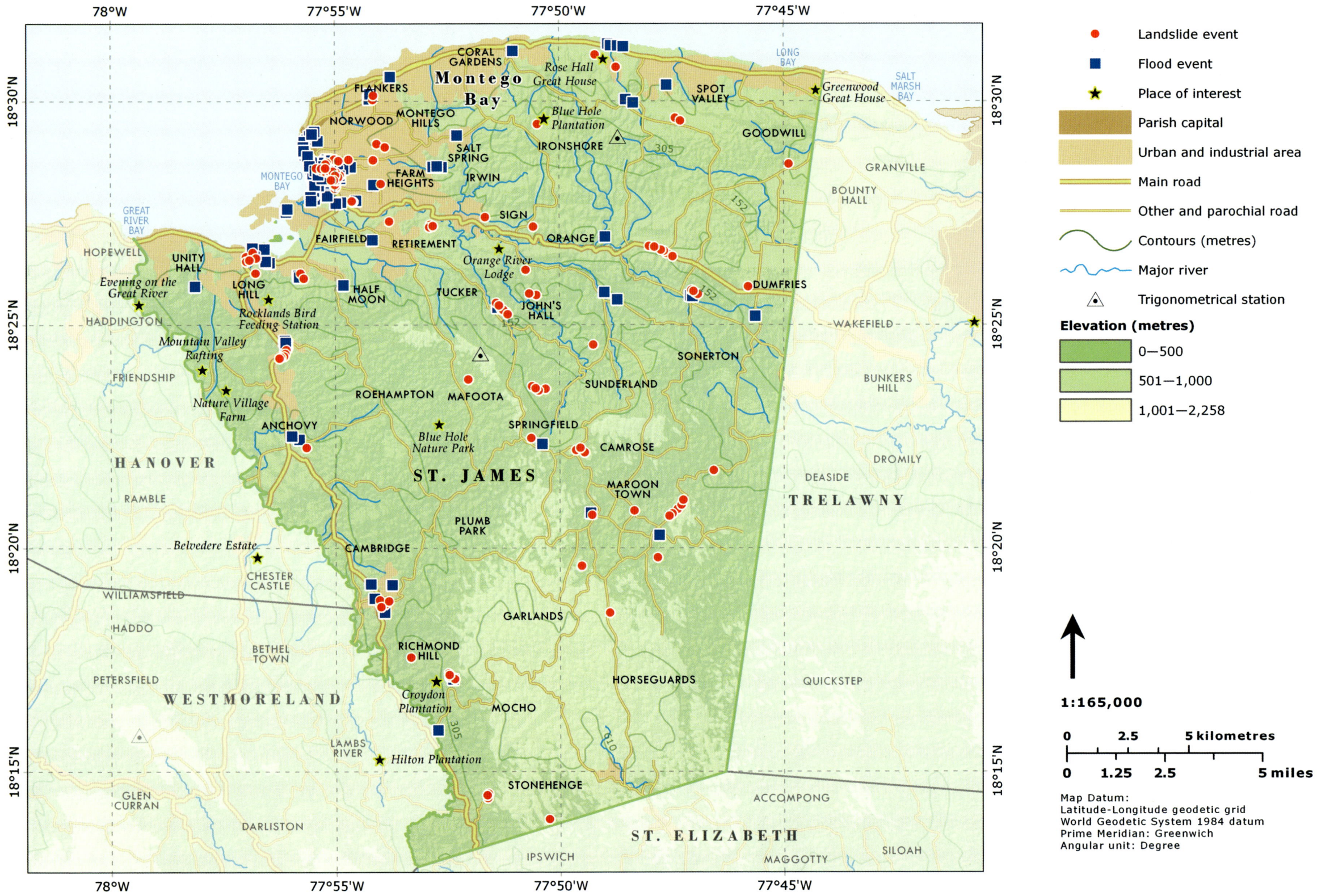

Map 3.8.1 St James landslide and flood distribution

Legend:
- ● Landslide event
- ■ Flood event
- ★ Place of interest
- Parish capital
- Urban and industrial area
- Main road
- Other and parochial road
- Contours (metres)
- Major river
- △ Trigonometrical station

Elevation (metres)
- 0—500
- 501—1,000
- 1,001—2,258

1:165,000

Map Datum:
Latitude-Longitude geodetic grid
World Geodetic System 1984 datum
Prime Meridian: Greenwich
Angular unit: Degree

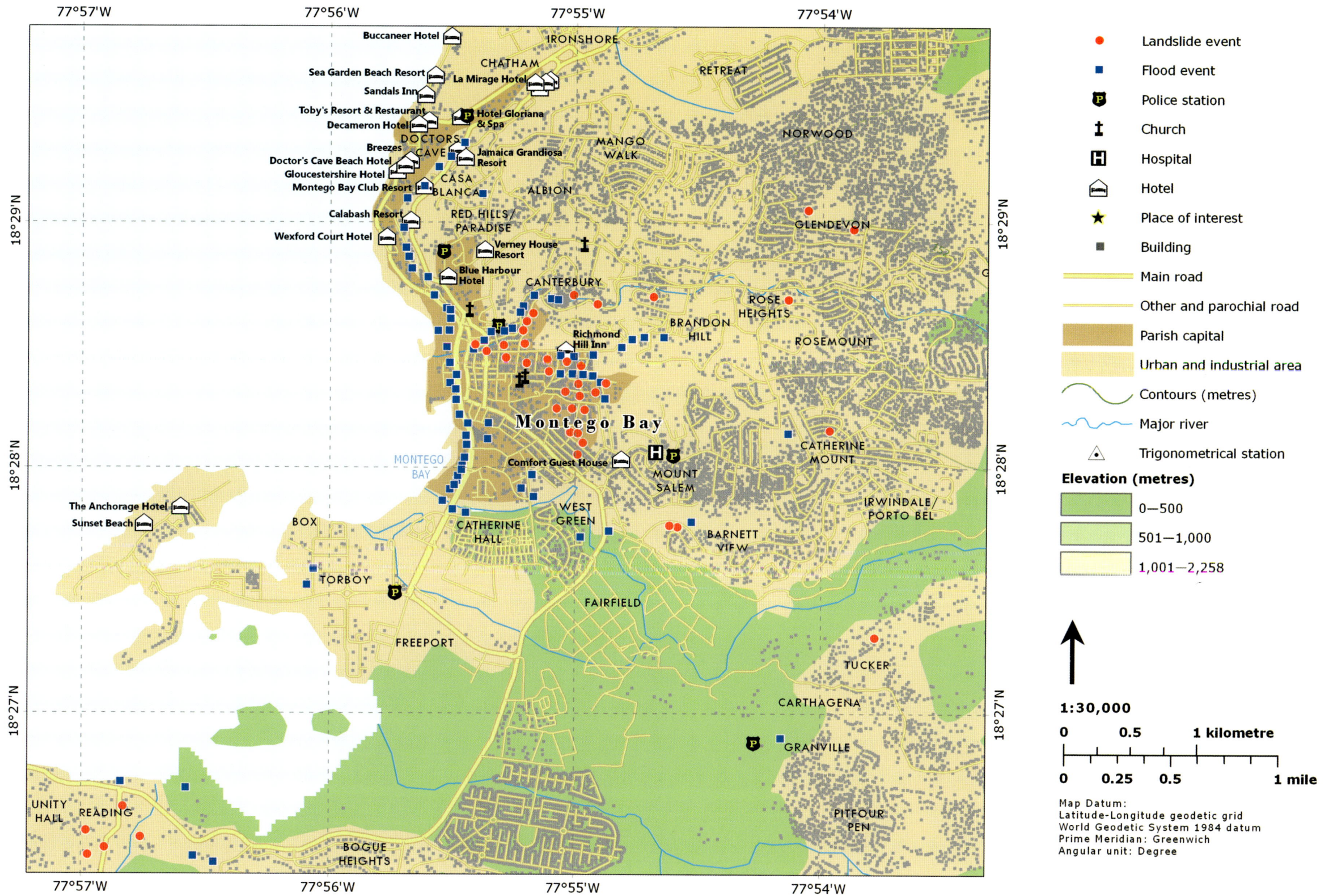

Map 3.8.2 Montego Bay landslide and flood distribution

Legend

- 🔴 Landslide event
- 🟦 Flood event
- Ⓟ Police station
- ✝ Church
- Ⓗ Hospital
- ⌂ Hotel
- ★ Place of interest
- ▪ Building
- Main road
- Other and parochial road
- Parish capital
- Urban and industrial area
- Contours (metres)
- Major river
- △ Trigonometrical station

Elevation (metres)

- 0—500
- 501—1,000
- 1,001—2,258

1:30,000

| 0 | 0.5 | 1 kilometre |

| 0 | 0.25 | 0.5 | 1 mile |

Map Datum:
Latitude-Longitude geodetic grid
World Geodetic System 1984 datum
Prime Meridian: Greenwich
Angular unit: Degree

3.9 HANOVER

Parish Profile

Human geography: The population of Hanover is just over 65,000. Main economic activities are agriculture and tourism, with significant expansion of the tourism-related infrastructure planned throughout the parish. The capital of Hanover is Lucea.

Physical geography: The parish has a varied landscape: bedrock comprises limestone and volcaniclastic sediments, and there are alluvial fans and marsh lands.

Hazard profile: Frequent flooding and landslides occur throughout the parish, especially inland.

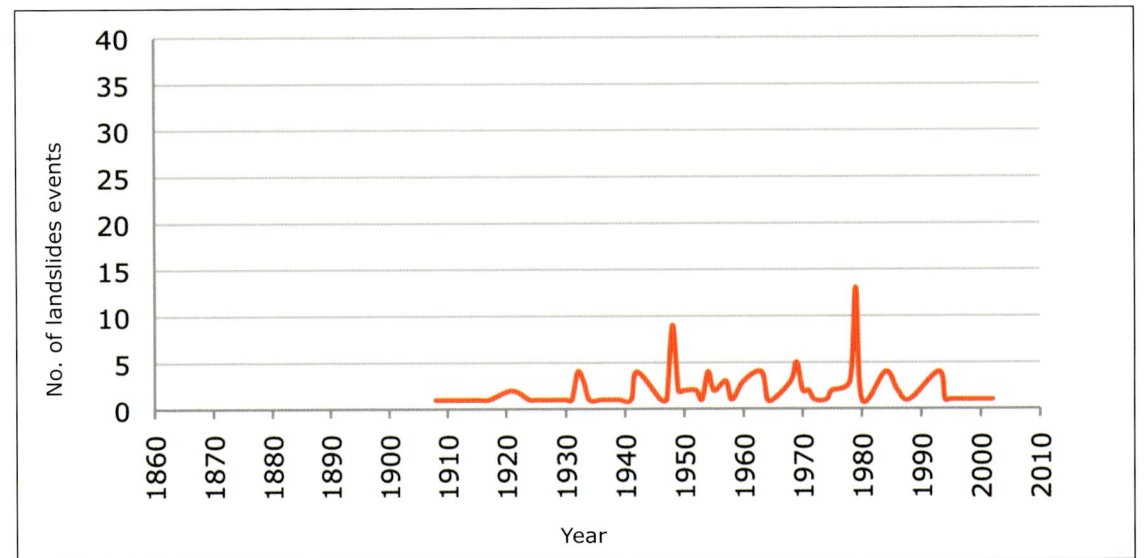

Graph 3.9.1 Hanover landslides

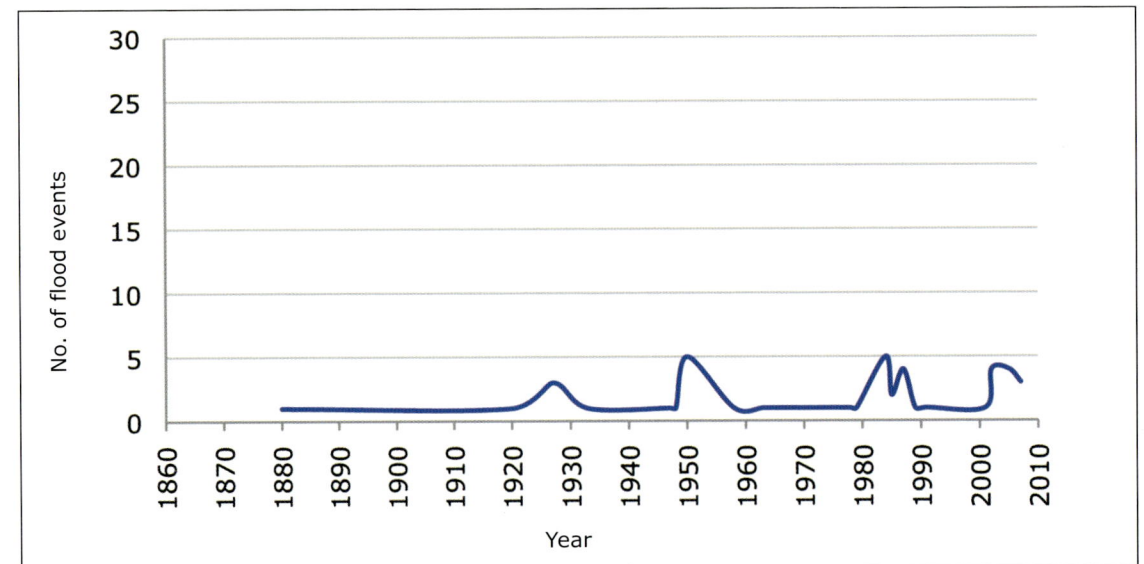

Graph 3.9.2 Hanover floods

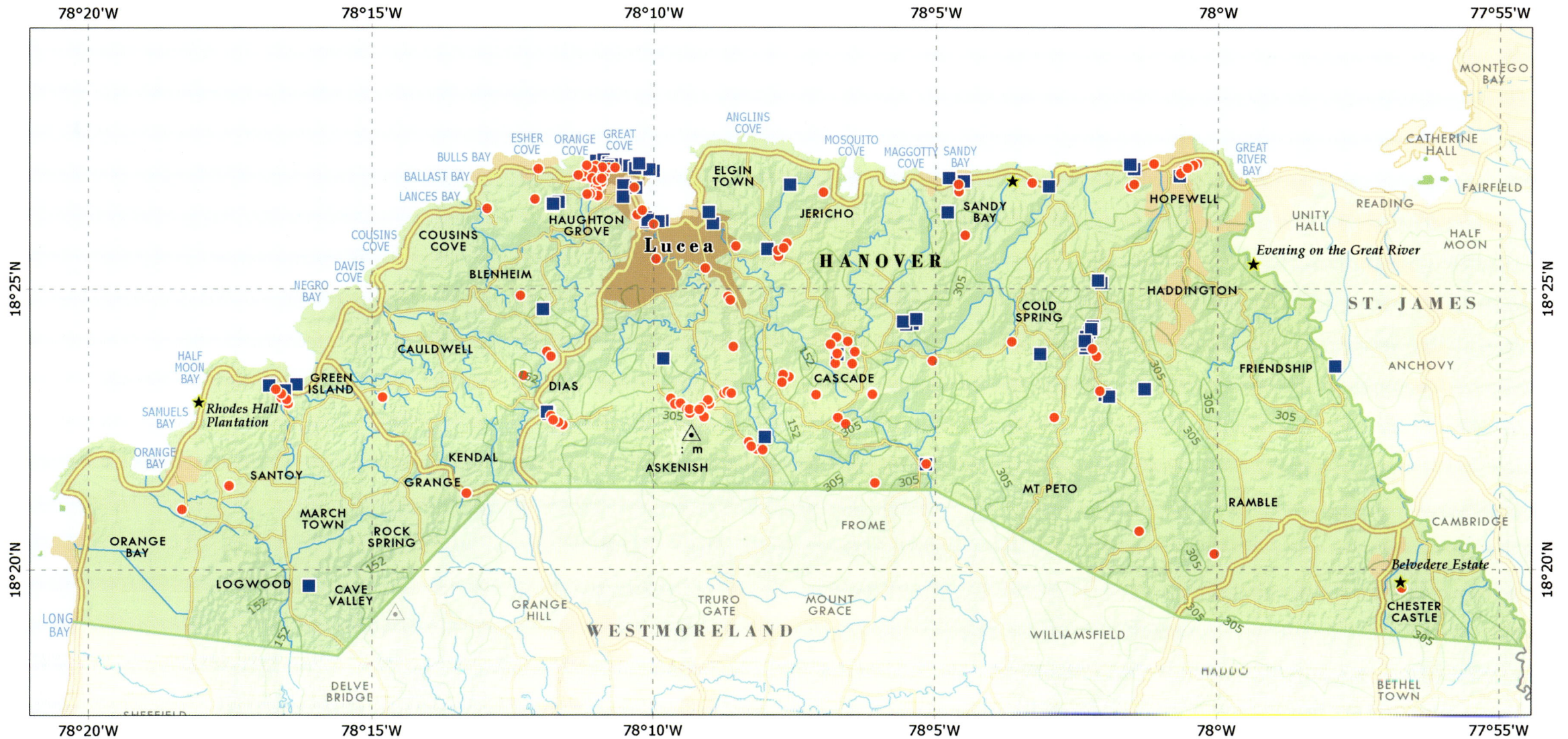

Legend

- ● Landslide event
- ■ Flood event
- ★ Place of interest
- ▨ Parish capital
- ▨ Urban and industrial area

- ▭ Main road
- ▭ Other and parochial road
- 〰 Contours (metres)
- 〰 Major river
- △ Trigonometrical station

Elevation (metres)
- 0—500
- 501—1,000
- 1,001—2,258

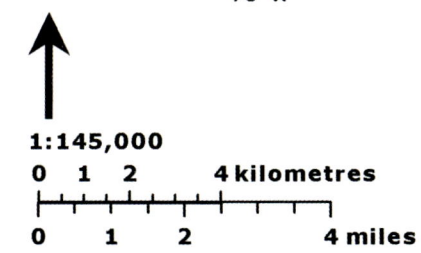

1:145,000

```
0   1   2        4 kilometres

0      1      2          4 miles
```

Map Datum:
Latitude-Longitude geodetic grid
World Geodetic System 1984 datum
Prime Meridian: Greenwich
Angular unit: Degree

Map 3.9.1 Hanover landslide and flood distribution

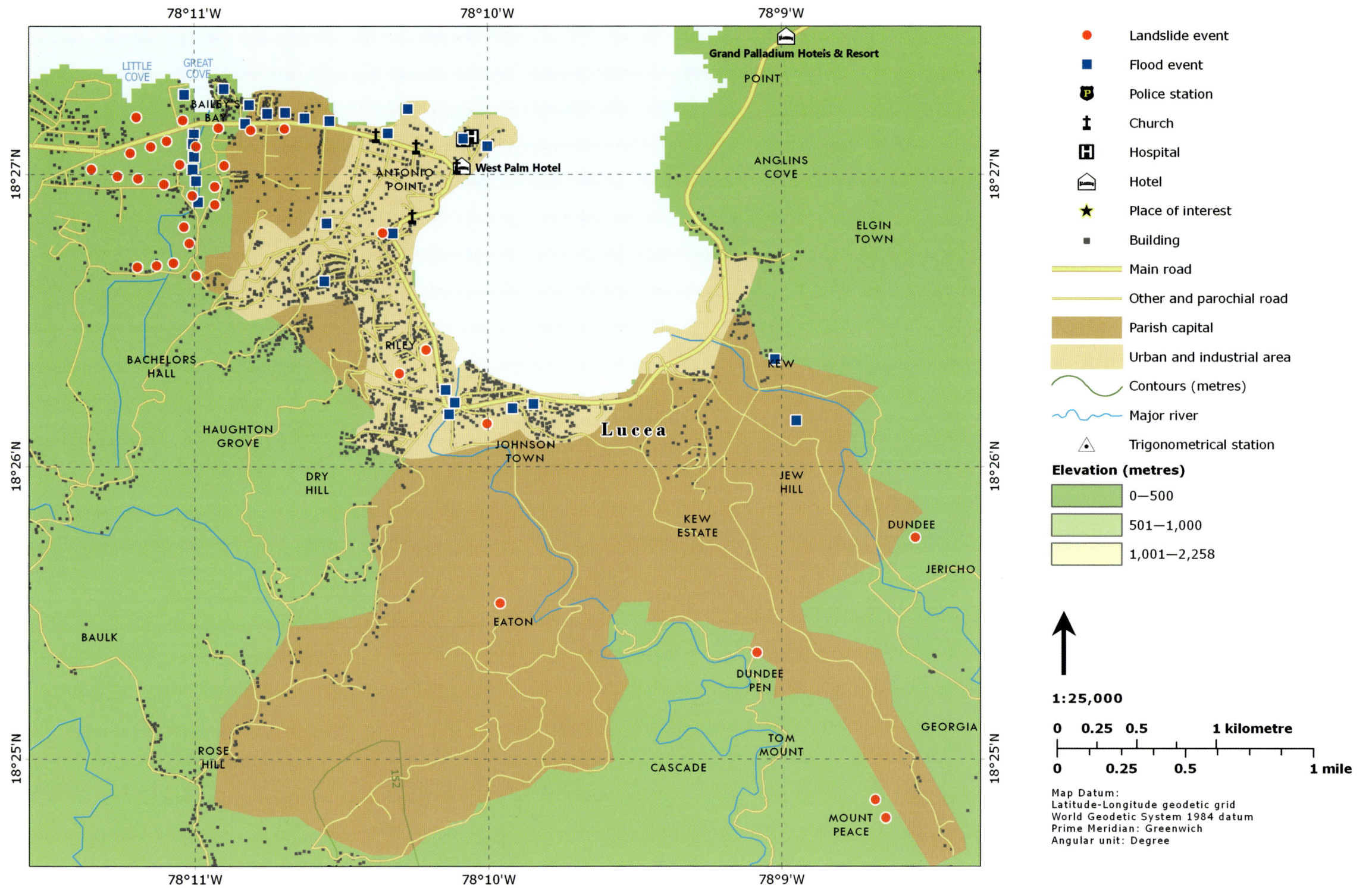

Map 3.9.2 Lucea landslide and flood distribution

Legend:

- ● Landslide event
- ■ Flood event
- Police station
- † Church
- Ⓗ Hospital
- Hotel
- ★ Place of interest
- ▪ Building
- Main road
- Other and parochial road
- Parish capital
- Urban and industrial area
- Contours (metres)
- Major river
- △ Trigonometrical station

Elevation (metres)

- 0—500
- 501—1,000
- 1,001—2,258

1:25,000

| 0 | 0.25 | 0.5 | | 1 kilometre |
| 0 | | 0.25 | 0.5 | 1 mile |

Map Datum:
Latitude-Longitude geodetic grid
World Geodetic System 1984 datum
Prime Meridian: Greenwich
Angular unit: Degree

Map labels: LITTLE COVE, GREAT COVE, BAILEY'S BAY, ANTONIO POINT, West Palm Hotel, Grand Palladium Hotels & Resort, POINT, ANGLINS COVE, ELGIN TOWN, KEW, BACHELORS HALL, HAUGHTON GROVE, RILEY, DRY HILL, JOHNSON TOWN, Lucea, KEW ESTATE, JEW HILL, DUNDEE, JERICHO, BAULK, EATON, DUNDEE PEN, TOM MOUNT, GEORGIA, ROSE HILL, CASCADE, MOUNT PEACE

3.10   WESTMORELAND

PARISH PROFILE

Human geography: Westmoreland has a population of just under 140,000 people. Main economic activities include agriculture (Jamaica's largest sugar estate, Frome, is in the parish), and tourism (particularly in Negril). The capital of Westmoreland is Savanna-la-Mar.

Physical geography: The parish has a varied landscape, with generally low-lying and heavily cultivated plains, limestone, volaniclastic rocks, marsh lands and peat deposits, and alluvial sediments.

Hazard profile: Westmoreland is highly susceptible to flooding and frequent landslides. Both Savanna-la-Mar and Negril, the two most important towns in the parish, are very susceptible to coastal flooding, including sea-level rise. Savanna-la-Mar, in 1912, experienced a dramatic storm surge, which carried an ocean-going vessel ashore. Debris flooding is also common, with a significant event occurring at Bluefields in 1979. Inland, flat areas used to cultivate sugar cane are also susceptible to flooding events.

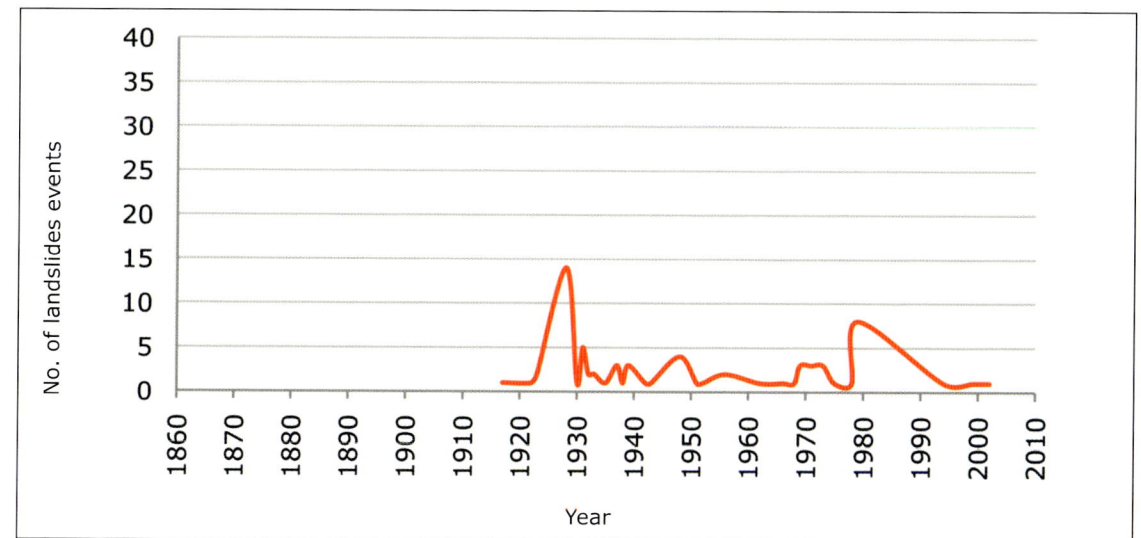

Graph 3.10.1 Westmoreland landslides

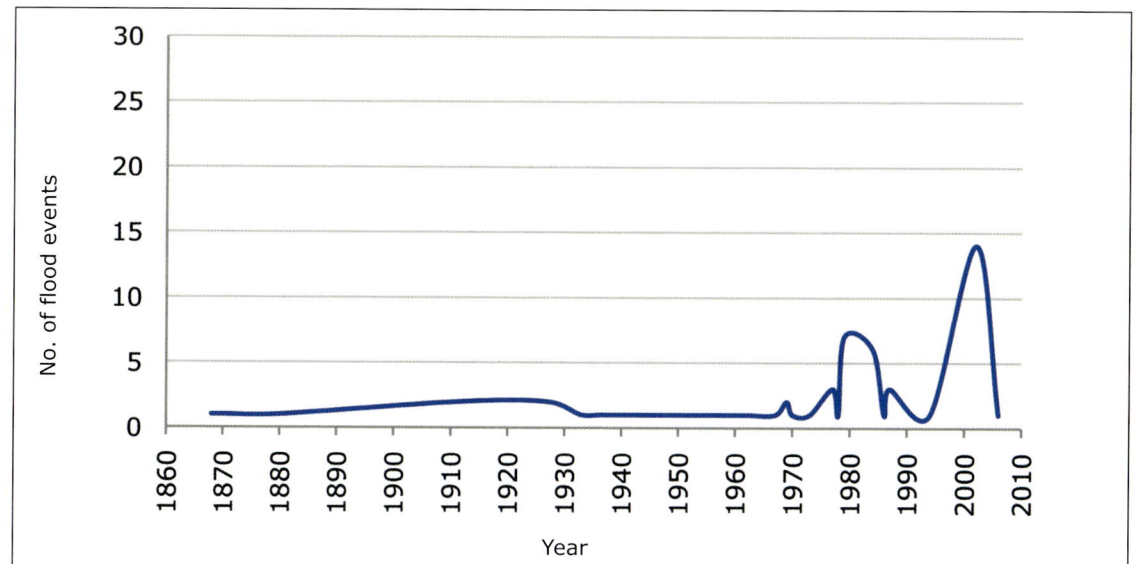

Graph 3.10.2 Westmoreland floods

Legend

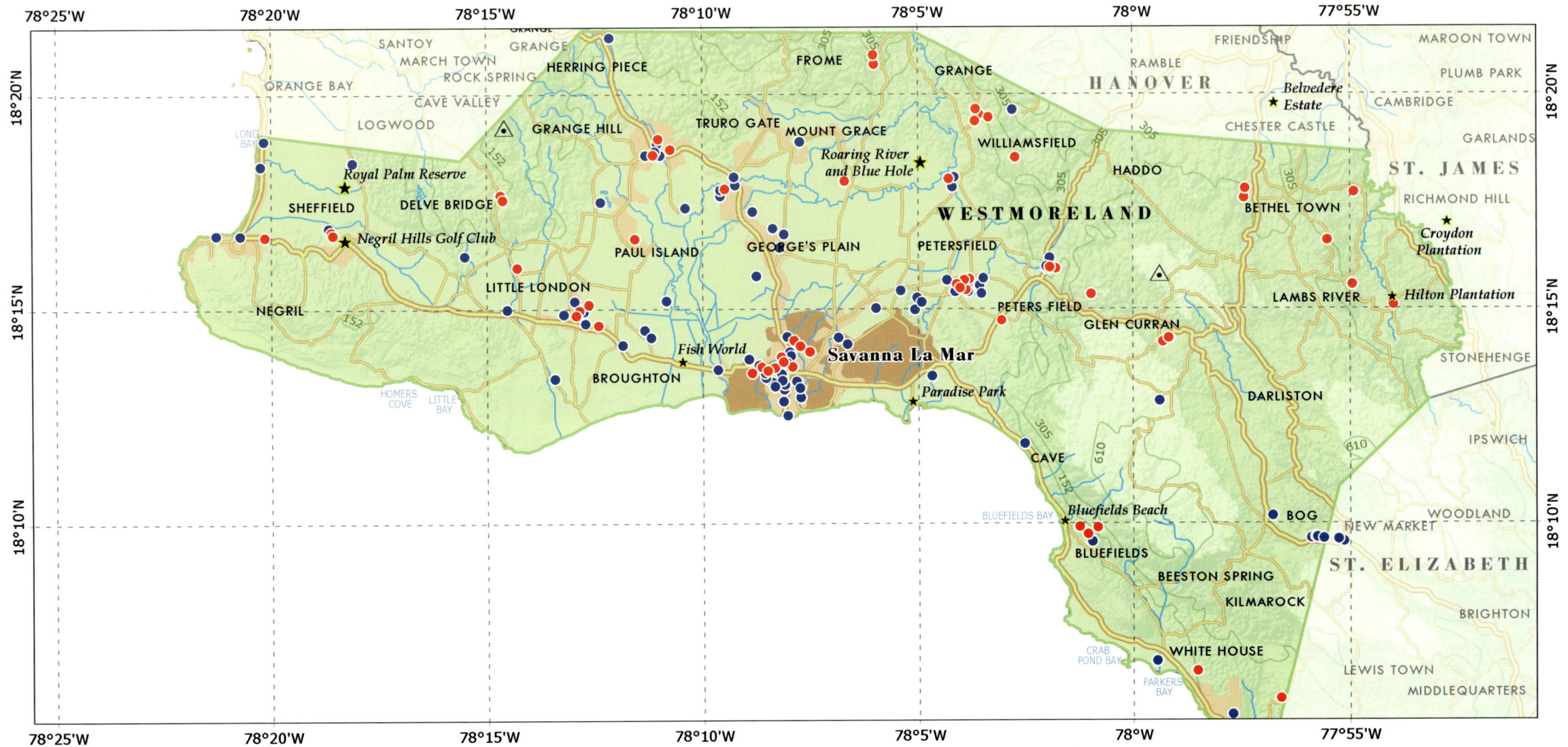

- ● Landslide event
- ● Flood event
- ★ Place of interest
- ▨ Parish capital
- ▨ Urban and industrial area

- Main road
- Other and parochial road
- Contours (metres)
- Major river
- △ Trigonometrical station

Elevation (metres)
- 0—500
- 501—1,000
- 1,001—2,258

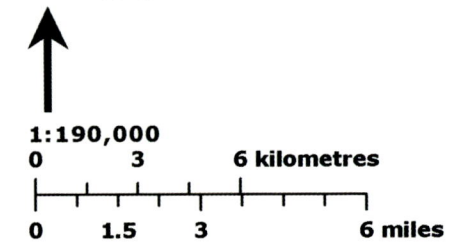

1:190,000

| 0 | 3 | 6 kilometres |

| 0 | 1.5 | 3 | 6 miles |

Map Datum:
Latitude-Longitude geodetic grid
World Geodetic System 1984 datum
Prime Meridian: Greenwich
Angular unit: Degree

Map 3.10.1 Westmoreland landslide and flood distribution

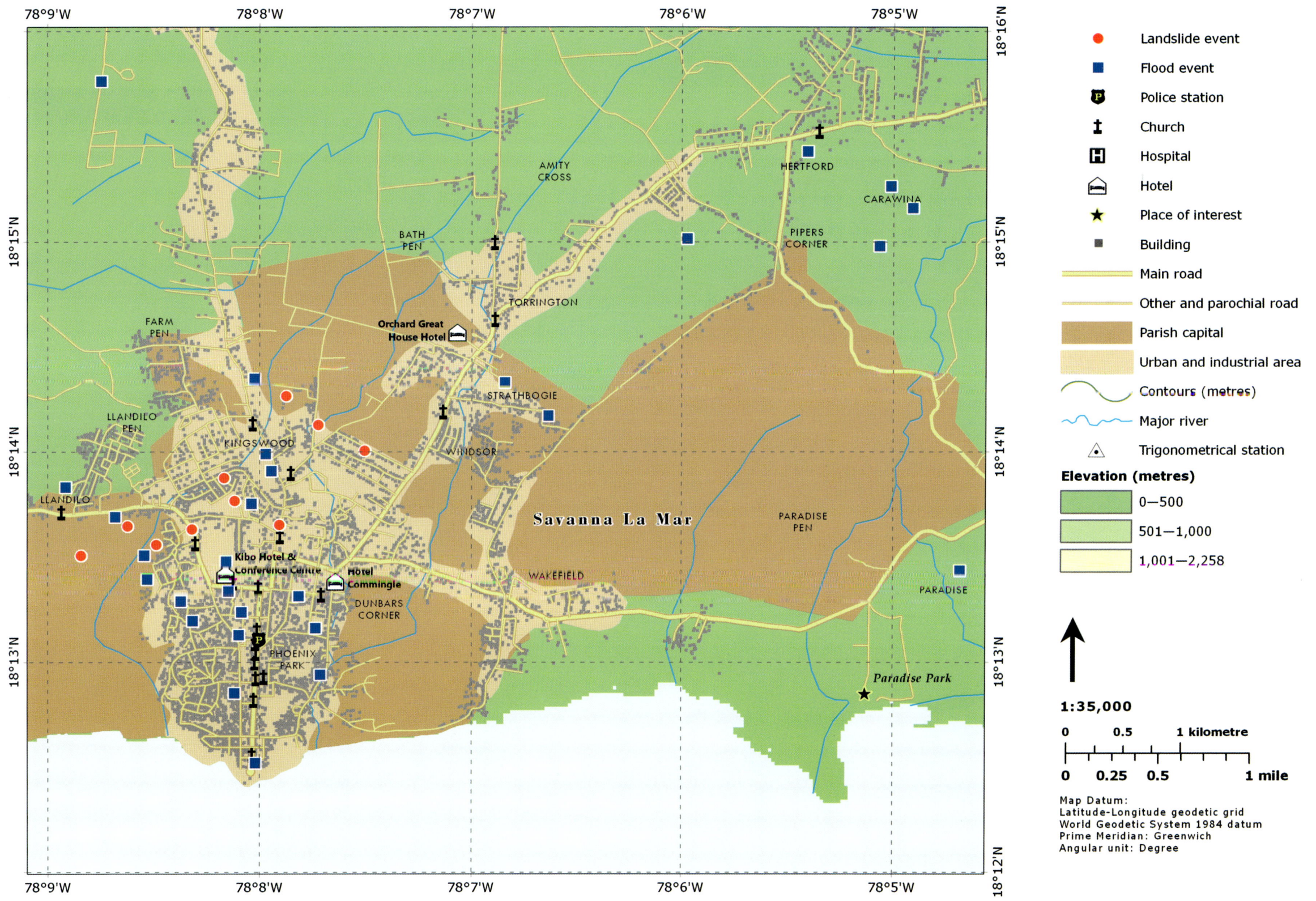

Map 3.10.2 Savanna-la-Mar landslide and flood distribution

Legend

- ● Landslide event
- ■ Flood event
- Police station
- Church
- H Hospital
- Hotel
- ★ Place of interest
- ■ Building
- Main road
- Other and parochial road
- Parish capital
- Urban and industrial area
- Contours (metres)
- Major river
- △ Trigonometrical station

Elevation (metres)

- 0—500
- 501—1,000
- 1,001—2,258

1:35,000

```
0        0.5        1 kilometre
0   0.25   0.5        1 mile
```

Map Datum:
Latitude-Longitude geodetic grid
World Geodetic System 1984 datum
Prime Meridian: Greenwich
Angular unit: Degree

Map labels

AMITY CROSS
HERTFORD
CARAWINA
BATH PEN
TORRINGTON
PIPERS CORNER
FARM PEN
Orchard Great House Hotel
STRATHBOGIE
LLANDILO PEN
KINGSWOOD
WINDSOR
Savanna La Mar
PARADISE PEN
LLANDILO
Kibo Hotel & Conference Centre
Hotel Commingle
DUNBARS CORNER
WAKEFIELD
PARADISE
PHOENIX PARK
Paradise Park

3.11 ST ELIZABETH

PARISH PROFILE

Human geography: St Elizabeth has a population of just under 150,000 people. The main economic activities are agriculture (sugar cane and small farming; St Elizabeth is often referred to as Jamaica's breadbasket), tourism, and bauxite and alumina production. The capital of St Elizabeth is Black River.

Physical geography: Cockpit Country is found to the north of the parish. It is drained by the Black River system; the upper Black River passes through the Nassau Valley polje, while the lower Black River passes through swamps to the south. The Santa Cruz Mountains lie to the south, and fault-controlled graben to the southeast.

Hazard profile: The parish is highly susceptible to flooding. The karst flood event at Newmarket in 1979 was a watershed in Jamaica's natural hazards history. The low-lying town of Black River, at the mouth of the river with the same name, is also prone to coastal flooding.

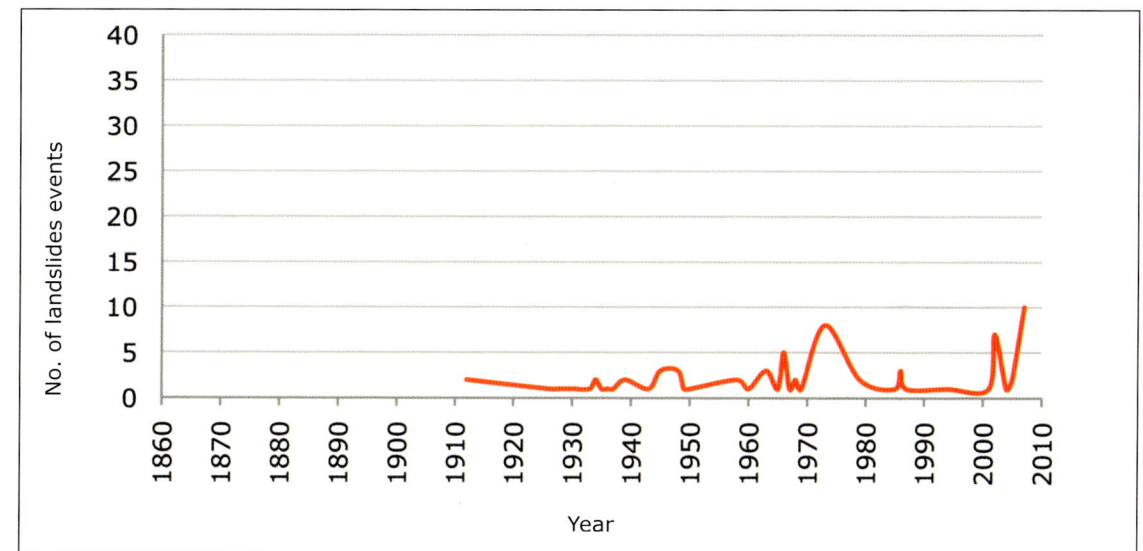

Graph 3.11.1 St Elizabeth landslides

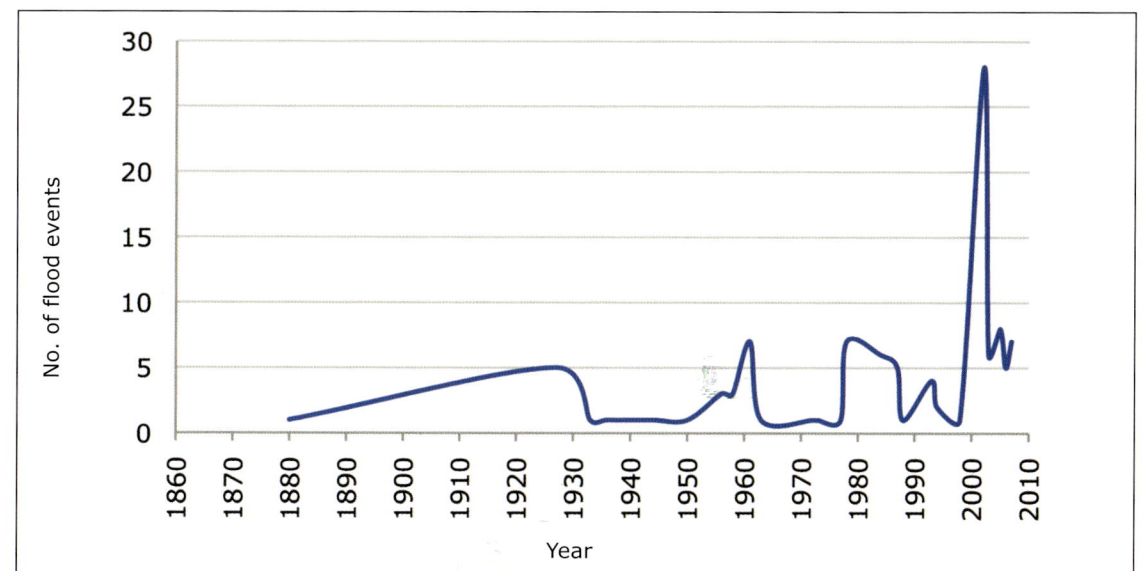

Graph 3.11.2 St Elizabeth floods

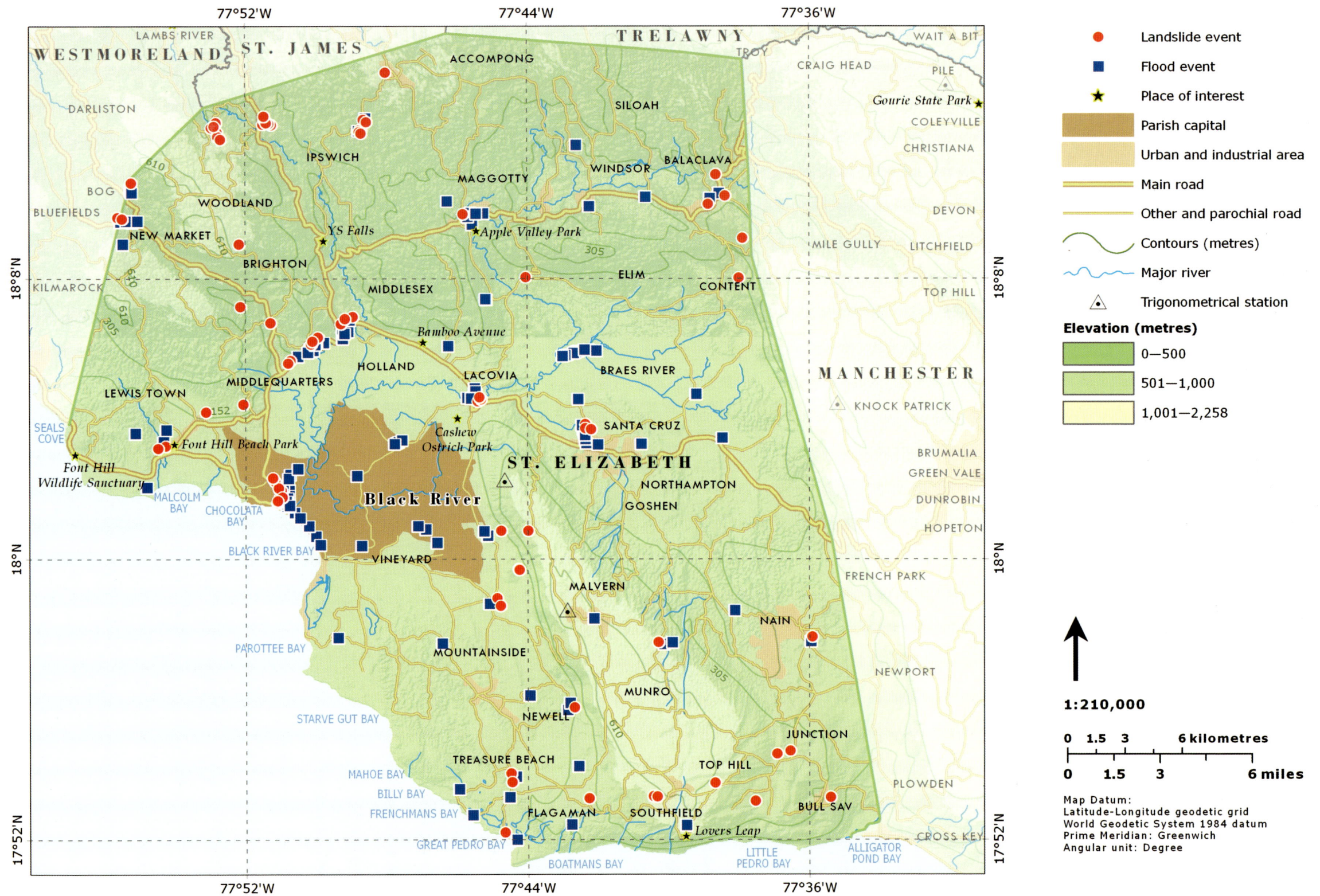

Map 3.11.1 St Elizabeth landslide and flood distribution

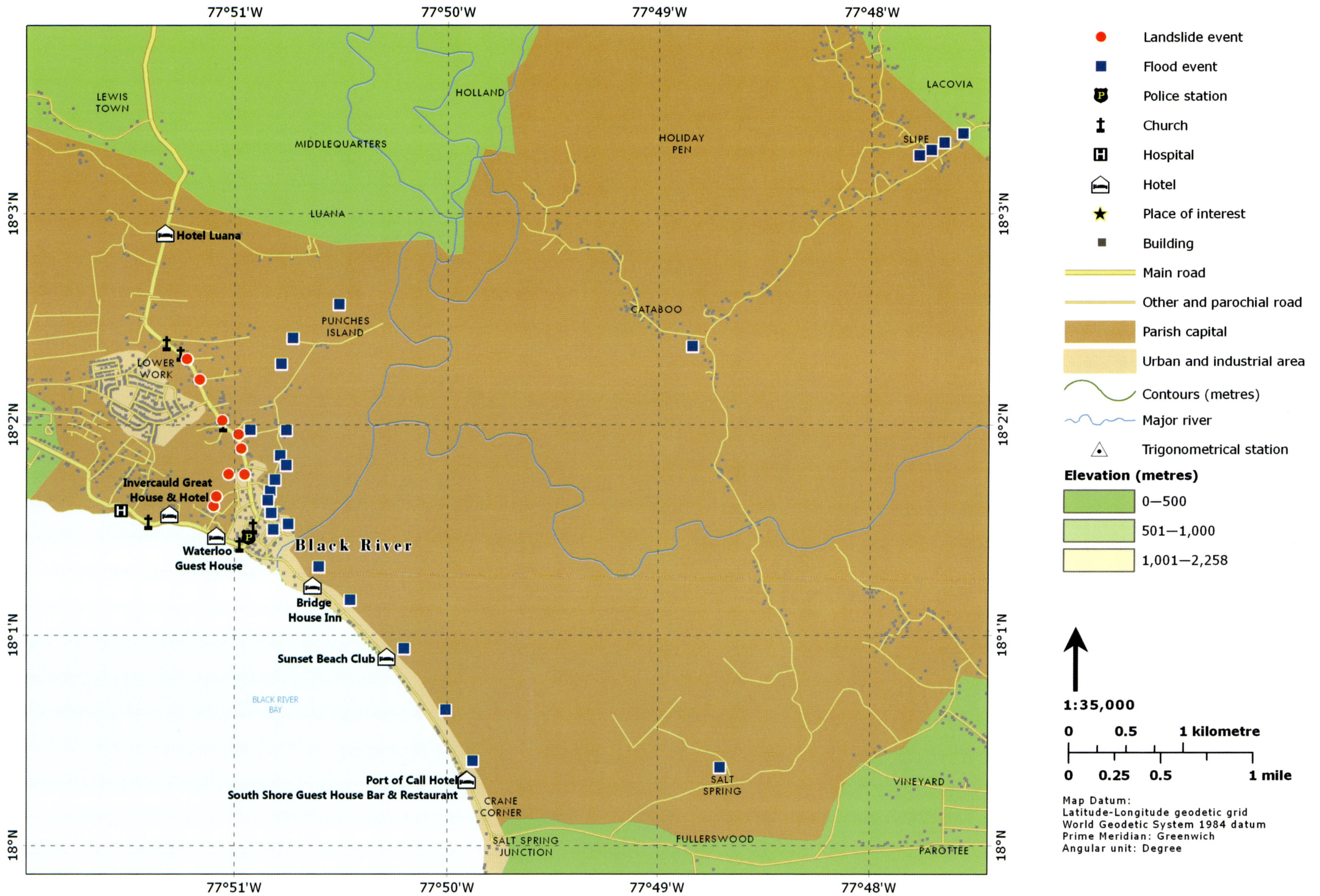

Legend

- ● Landslide event
- ■ Flood event
- Ⓟ Police station
- ✝ Church
- Ⓗ Hospital
- ⌂ Hotel
- ★ Place of interest
- ■ Building
- Main road
- Other and parochial road
- Parish capital
- Urban and industrial area
- Contours (metres)
- Major river
- △ Trigonometrical station

Elevation (metres)

- 0—500
- 501—1,000
- 1,001—2,258

1:35,000

0   0.5   1 kilometre

0   0.25   0.5   1 mile

Map Datum:
Latitude-Longitude geodetic grid
World Geodetic System 1984 datum
Prime Meridian: Greenwich
Angular unit: Degree

Map 3.11.2 Black River landslide and flood distribution

Plate 3.11.1 Newmarket, flood rains, October 2002 (Water Resources Authority, 2002)

Plate 3.11.2 Newmarket, flooded by rainfall associated with Tropical Depression Nicole, September–October 2010 (R. Ahmad, 2010)

Plate 3.11.3 Newmarket road, remains submerged in December 2010 (R. Ahmad, 2010)

3.12 MANCHESTER

PARISH PROFILE

Human geography: The population of Manchester is just over 185,000 people. Main economic activities include agriculture (mostly small farming), and bauxite and alumina. The capital of Manchester is Mandeville.

Physical geography: The parish is largely an uplifted plateau. Like St Ann, Manchester has a small portion of its elevation close to sea level; most of the parish is at higher elevations. There are few river systems in Manchester, which result in significant water shortages for residents.

Hazard profile: Manchester is prone to significant karst flooding, especially in areas such as Harmons and Porus. Manchester is also susceptible to high wind speeds, owing to its elevation and gently varying topography.

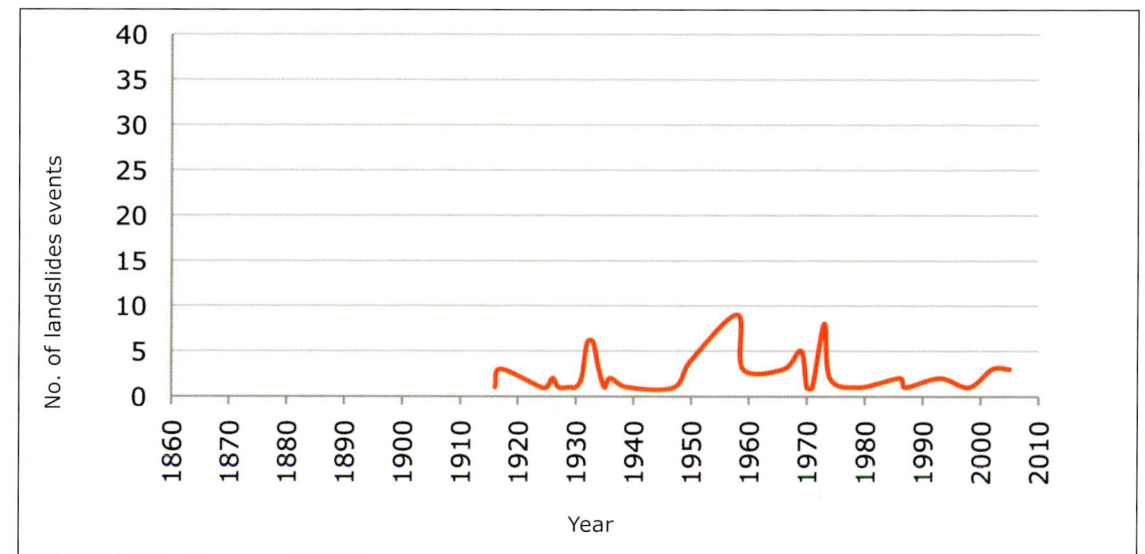

Graph 3.12.1 Manchester landslides

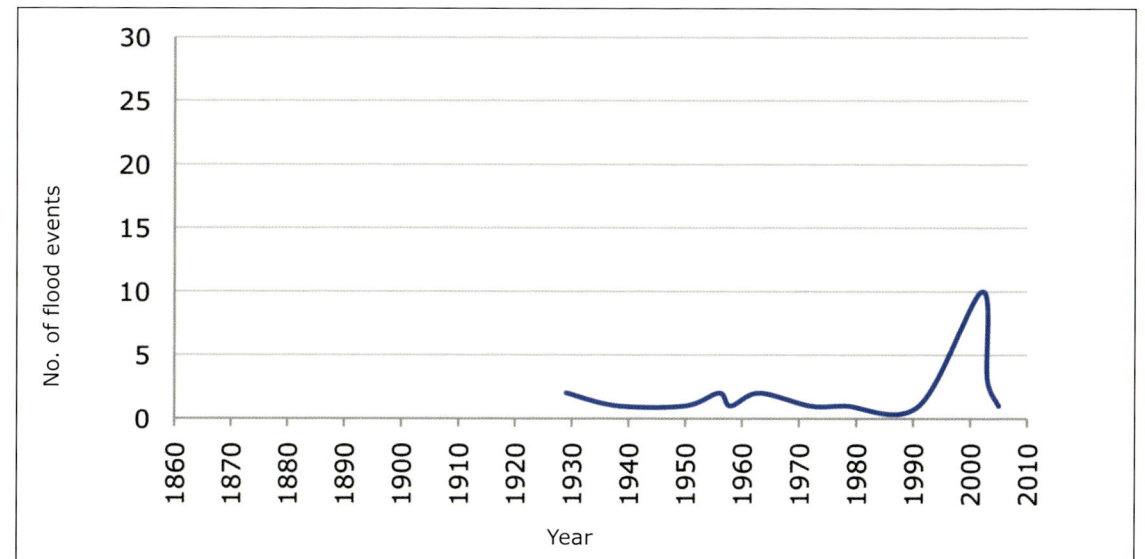

Graph 3.12.2 Manchester floods

Landslide event
Flood event
Place of interest
Parish capital
Urban and industrial area
Main road
Other and parochial road
Contours (metres)
Major river
Trigonometrical station

Elevation (metres)

0—500
501—1,000
1,001—2,258

TRELAWNY   WARSOP   LITCHFIELD   ST. ANN

ACCOMPONG   CRAIG HEAD   WAIT A BIT

SILOAH   PILE   914   Gourie State Park   AENON TOWN

COLEYVILLE   ALSTON

MAGGOTTY   WINDSOR   BALACLAVA   CHRISTIANA   JAMES HILL
                         DEVON   KELLITS

Apple Valley Park   610   Christiana Bottom

ELIM   MILE GULLY   CHUDLEIGH   RICHES   CROOKED RIVER
CONTENT   LITCHFIELD   SPALDING   FRANKFIELD

18°8'N   18°8'N

LACOVIA   BRAES RIVER   ALLISON   PENNANTS
                         TOP HILL
Cashew Ostrich Park   MIZPATH   THOMPSON TOWN
              610
         SANTA CRUZ   KNOCK PATRICK   WILLIAMSFIELD   BECKFORD KRAAL
              High Mountain Coffee Factory
ST. ELIZABETH   Marshall's Pen   CLARKS TOWN   CLARENDON   MAY PEN
         NORTHAMPTON   RICHMOND PARK
BLACK RIVER   GOSHEN   BRUMALIA   MOCHO
                   Mandeville
                   HATFIELD   Huntington Summit   PORUS
18°N   18°N
         MALVERN   ALBION
                   HILLSIDE   OSBOURNE STORE
                   FRENCH PARK   FOUR PATHS
MOUNTAINSIDE   NAIN
         MUNRO
                   NEWPORT   305   YORK TOWN
NEWELL
         JUNCTION   PRATVILLE
TREASURE BEACH   TOP HILL   CROSS KEYS   VICTORIA TOWN
              BULL SAV   MANCHESTER   RACE COURSE
         PLOWDEN   RESOURCE   Alligator Hole
FLAGAMAN   SOUTHFIELD   152
         Lovers Leap   MILK RIVER
17°52'N   GREAT   Milk River Spa   17°52'N
         PEDRO BAY   CALABASH BAY
              GREEN BAY

1:220,000

0   4   8 kilometres
0   2   4   8 miles

Map Datum:
Latitude-Longitude geodetic grid
World Geodetic System 1984 datum
Prime Meridian: Greenwich
Angular unit: Degree

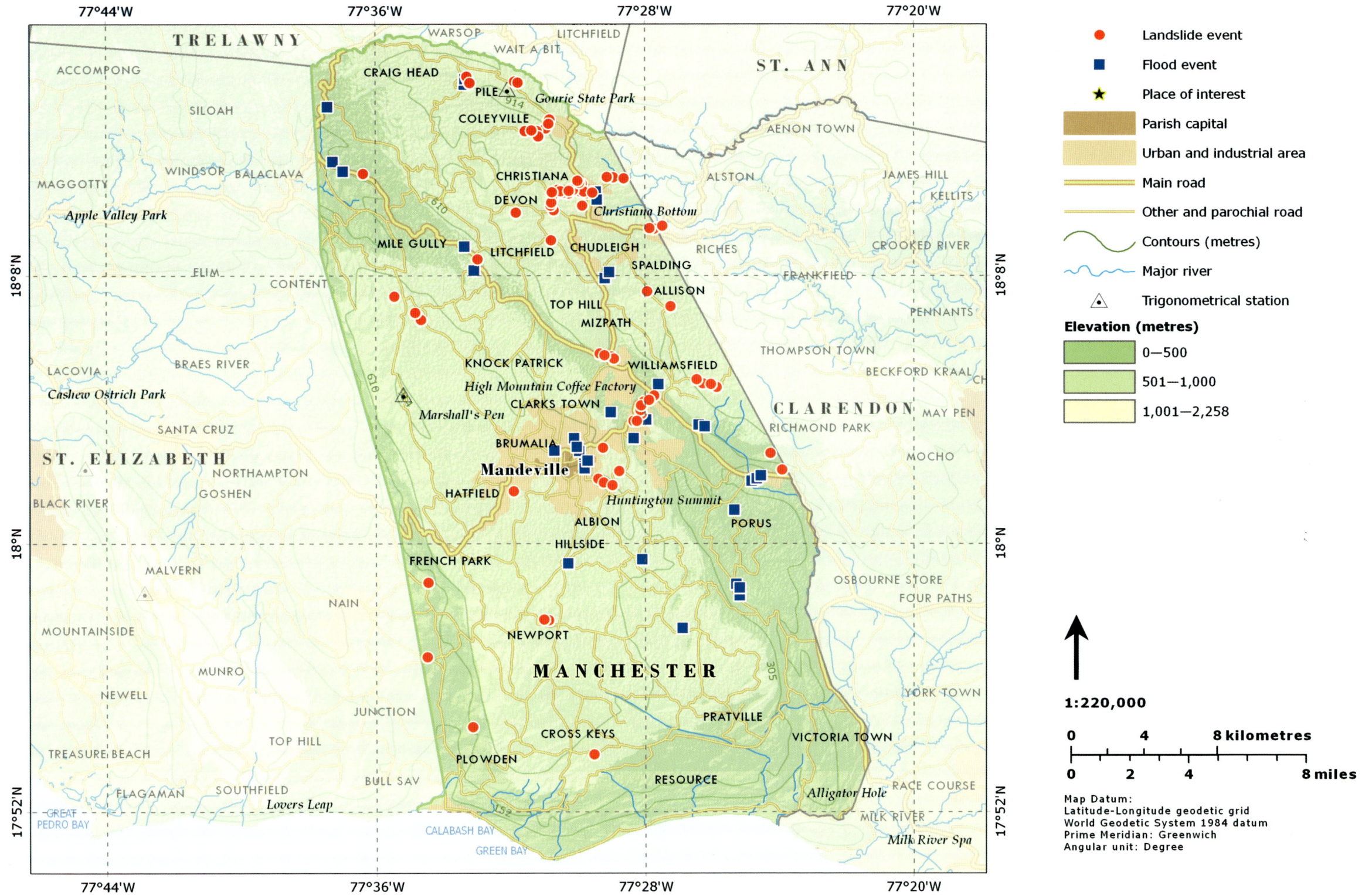

Map 3.12.1 Manchester landslide and flood distribution

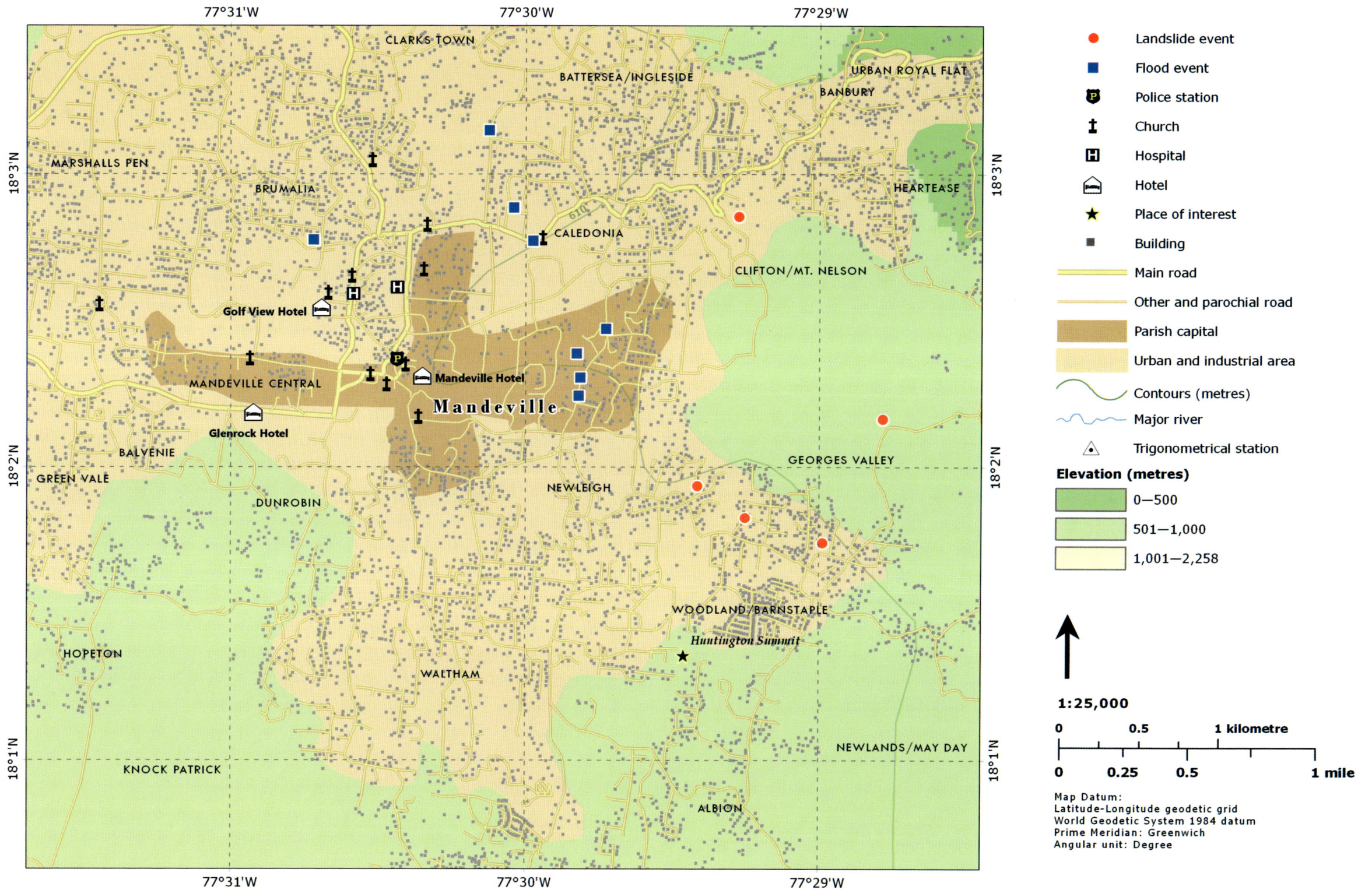

Map 3.12.2 Mandeville landslide and flood distribution

Legend:
- Landslide event
- Flood event
- P Police station
- Church
- H Hospital
- Hotel
- Place of interest
- Building
- Main road
- Other and parochial road
- Parish capital
- Urban and industrial area
- Contours (metres)
- Major river
- Trigonometrical station

Elevation (metres)
- 0—500
- 501—1,000
- 1,001—2,258

1:25,000

Map Datum:
Latitude-Longitude geodetic grid
World Geodetic System 1984 datum
Prime Meridian: Greenwich
Angular unit: Degree

3.13 CLARENDON

PARISH PROFILE

Human geography: The population of Clarendon is just under 250,000 people. The main economic activities include agriculture (sugar cane and small farming), and bauxite and alumina production. The capital of Clarendon is May Pen.

Physical geography: The uplands of the parish have deeply weathered volcaniclastic sediments, volcanic rocks and limestone, which form the Central Inlier. The Rio Minho system drains to the south, creating the heavily cultivated Vere Plain.

Hazard profile: Clarendon is highly susceptible to flooding along the Vere Plain, associated with the Rio Minho and Milk River systems, and frequent landslides to the north. Many urban developments in the parish exacerbate the flooding problem in the parish by not taking into account surface drainage issues.

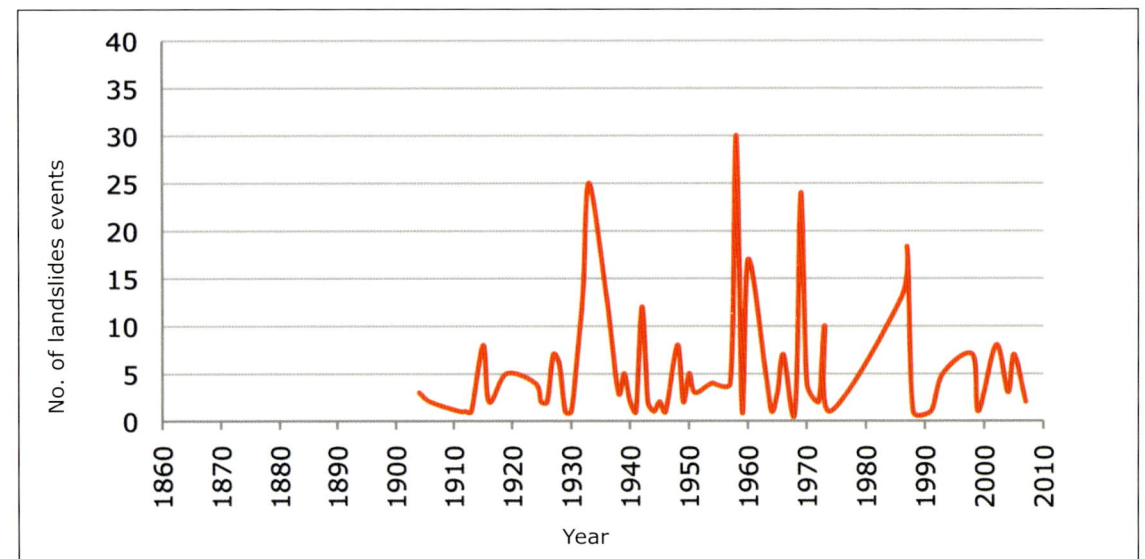

Graph 3.13.1 Clarendon landslides

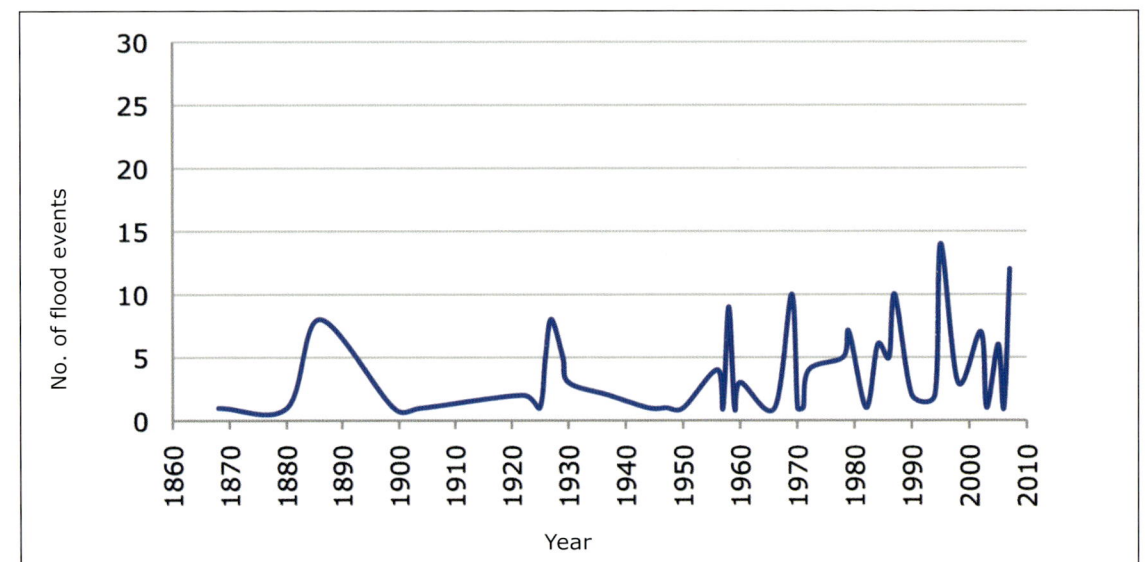

Graph 3.13.2 Clarendon floods

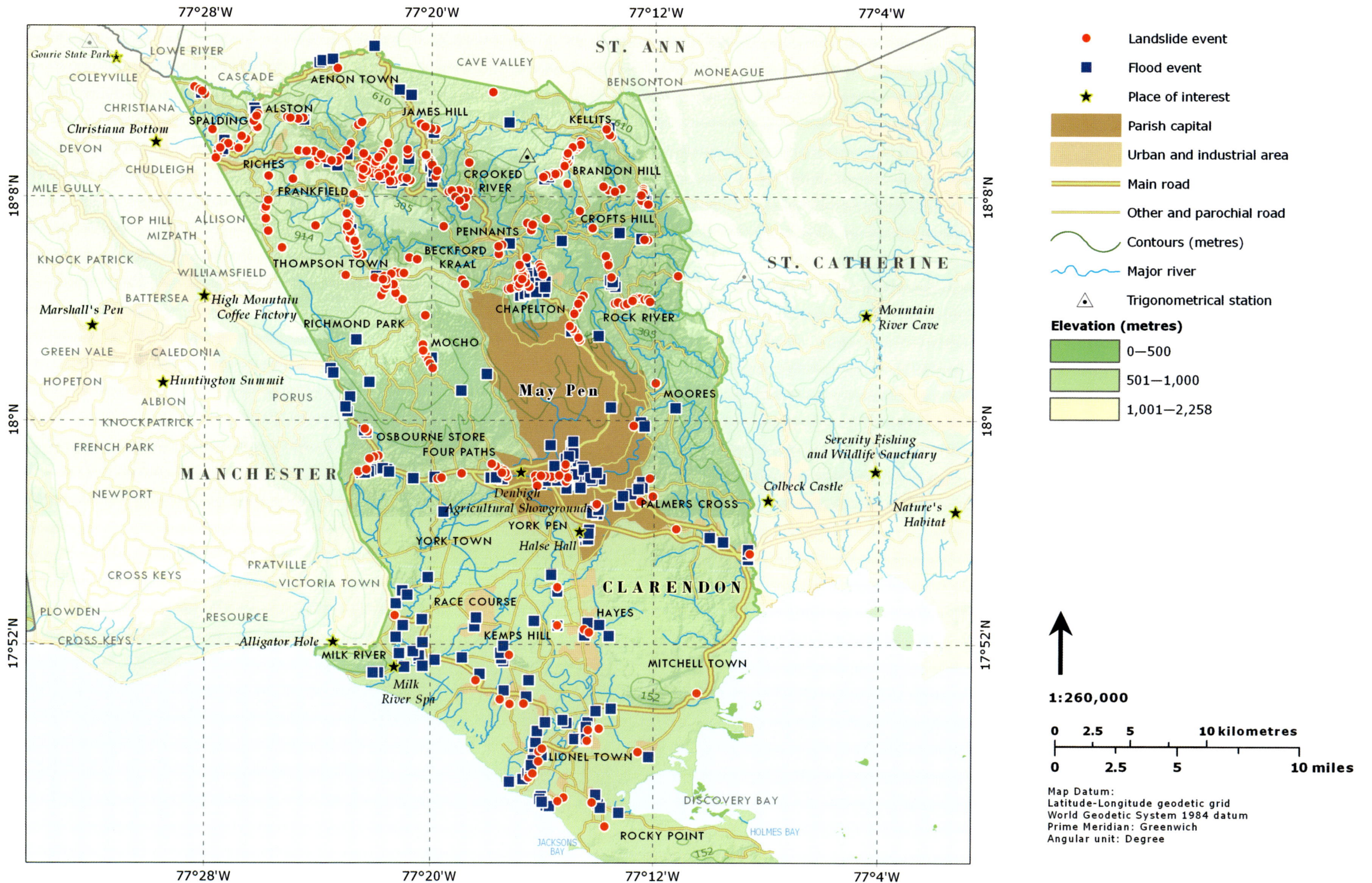

Map 3.13.1 Clarendon landslide and flood distribution

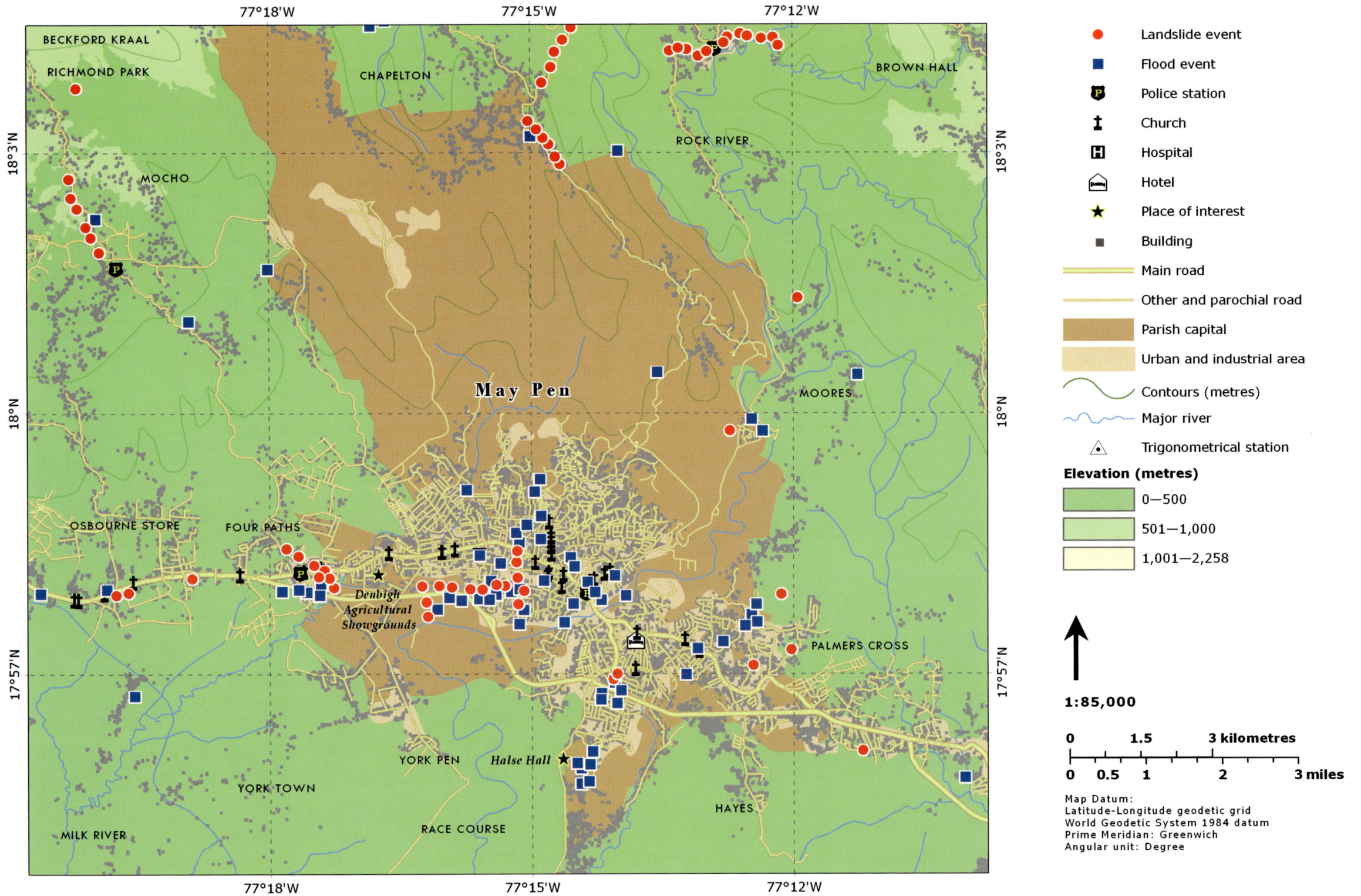

Map 3.13.2 May Pen landslide and flood distribution

Legend

- ● Landslide event
- ■ Flood event
- Ⓟ Police station
- ♰ Church
- Ⓗ Hospital
- ⌂ Hotel
- ★ Place of interest
- ▪ Building
- ━━ Main road
- ── Other and parochial road
- Parish capital
- Urban and industrial area
- ∿ Contours (metres)
- ∿ Major river
- △ Trigonometrical station

Elevation (metres)

- 0—500
- 501—1,000
- 1,001—2,258

1:85,000

| 0 | 1.5 | 3 kilometres |

| 0 | 0.5 | 1 | 2 | 3 miles |

Map Datum:
Latitude-Longitude geodetic grid
World Geodetic System 1984 datum
Prime Meridian: Greenwich
Angular unit: Degree

3.14 ST CATHERINE

PARISH PROFILE

Human geography: The population of St Catherine is just under 500,000 people. Agriculture (sugar cane and citrus farming) and bauxite and alumina production are the significant economic activities in the parish. The capital of St Catherine is Jamaica's original capital, Spanish Town, while the dormitory suburb of Portmore is the second-largest settlement in Jamaica and the most densely populated town in the Caribbean.

Physical geography: The parish has a varied landscape, with deeply weathered rocks and limestone in the uplands. The Rio Cobre system drains to the south dissecting the Bog Walk Gorge, emptying on to the Caymanas Plain. Karstic Hellshire Hills have little drainage. Heavily urbanized Portmore is developed on reclaimed swamp land to the southeast.

Hazard profile: St Catherine is highly susceptible to flooding and frequent landslides. The Bog Walk Gorge is frequently flooded during times of heavy rainfall, often resulting in significant damage to the main north-south corridor in Jamaica, as well as disruption of travel. Portmore is vulnerable to coastal flooding as well. St Catherine has had the most number of reported flood events, based on the *Gleaner* archives. Landslides are common further inland, particularly in the western upland parts of the parish. Earthquake-induced liquefaction may threaten structures in the reclaimed portions of Portmore.

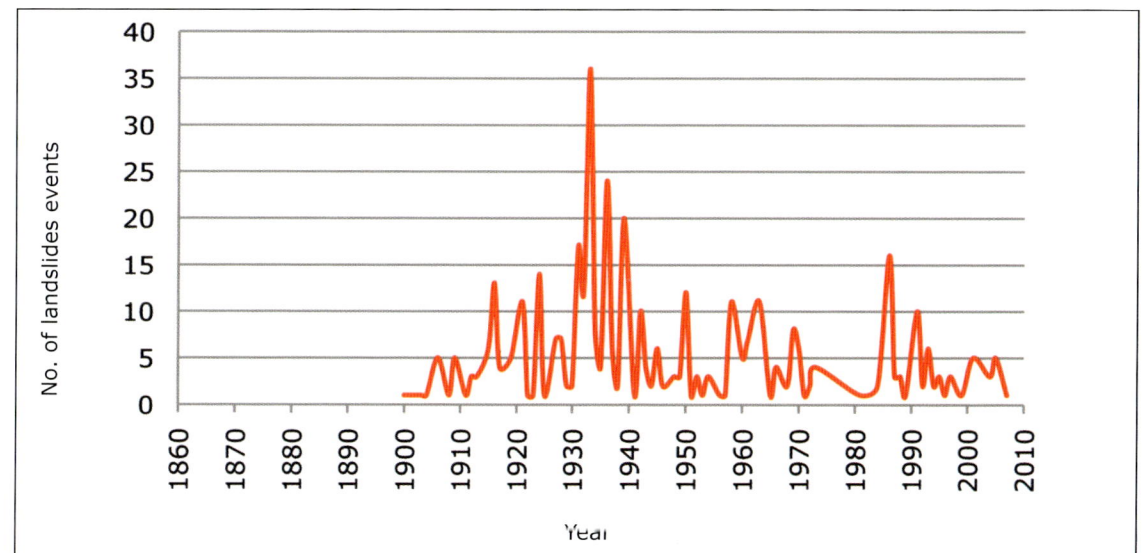

Graph 3.14.1 St Catherine landslides

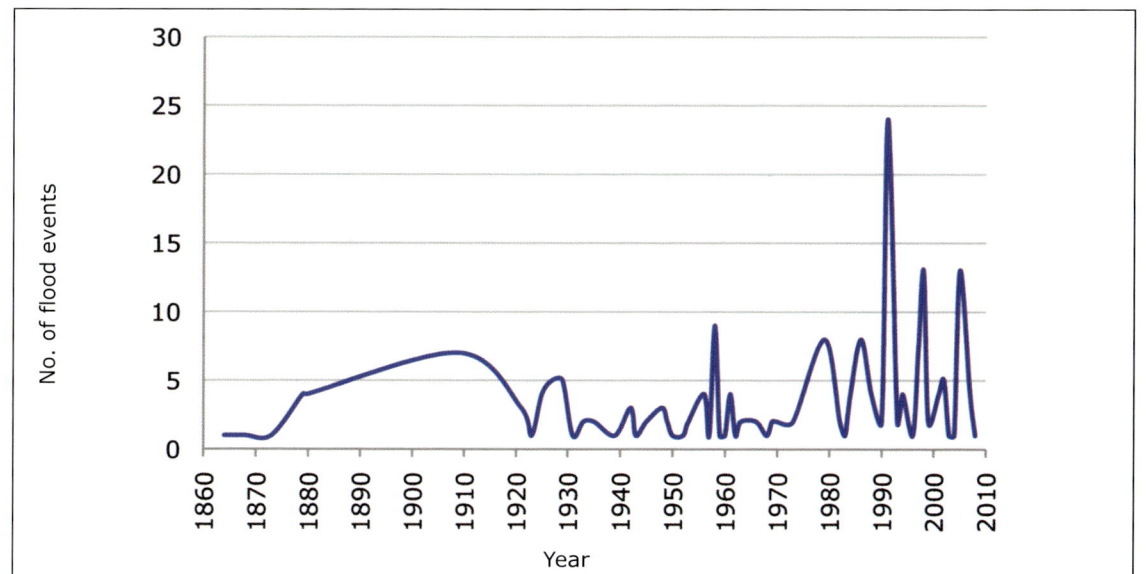

Graph 3.14.2 St Catherine floods

Plate 3.14.1 Damage to the Old Harbour Bay community, Hurricane Dean (P. Lyew-Ayee Jr, 2007)

Plate 3.14.2 Damage to the Flat Bridge, Tropical Storm Gustav (Water Resources Authority, 2008)

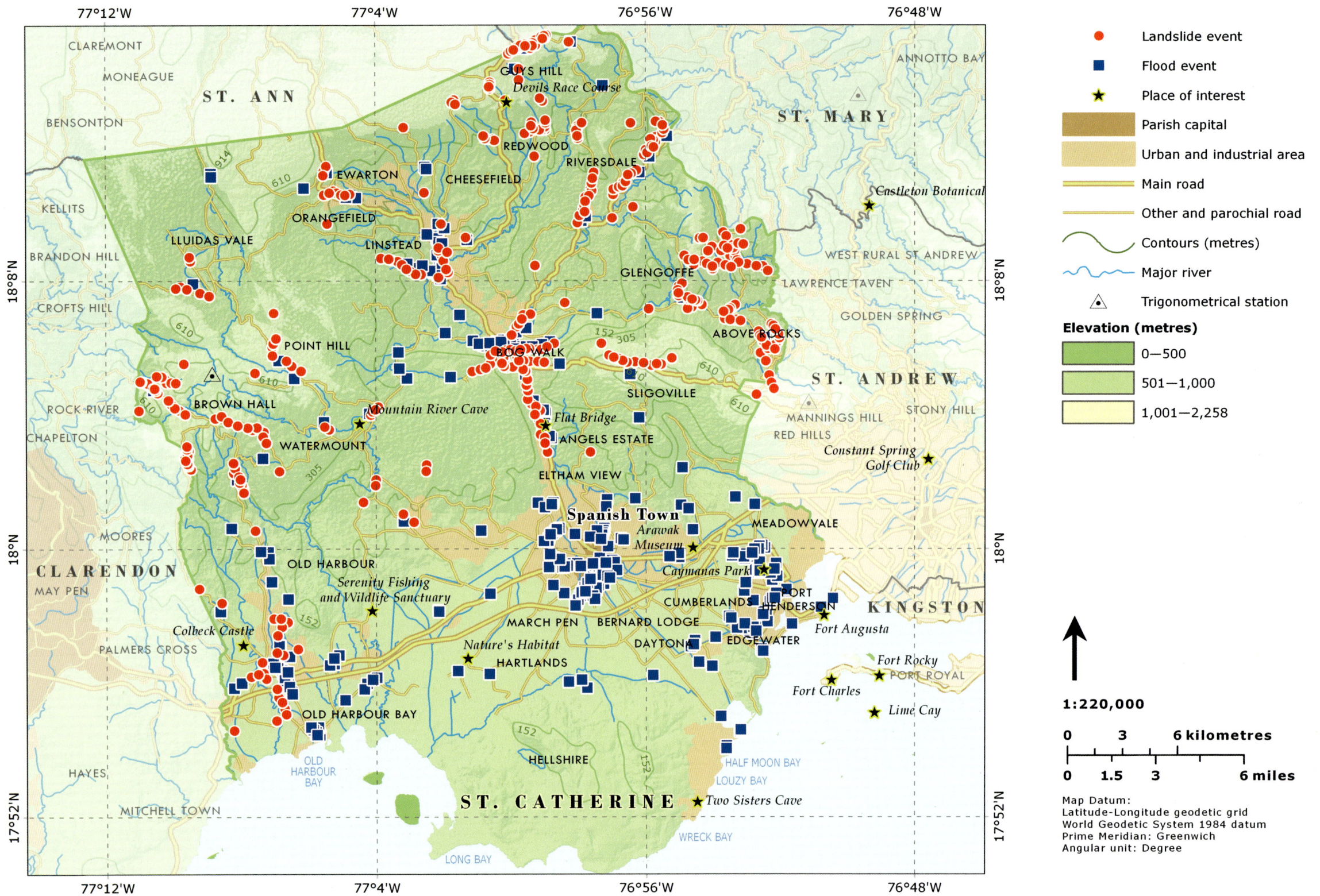

Map 3.14.1 St Catherine landslide and flood distribution

Legend:
- ● Landslide event
- ■ Flood event
- ★ Place of interest
- Parish capital
- Urban and industrial area
- Main road
- Other and parochial road
- Contours (metres)
- Major river
- ⬡ Trigonometrical station

Elevation (metres)
- 0–500
- 501–1,000
- 1,001–2,258

1:220,000

0   3   6 kilometres
0   1.5   3   6 miles

Map Datum:
Latitude-Longitude geodetic grid
World Geodetic System 1984 datum
Prime Meridian: Greenwich
Angular unit: Degree

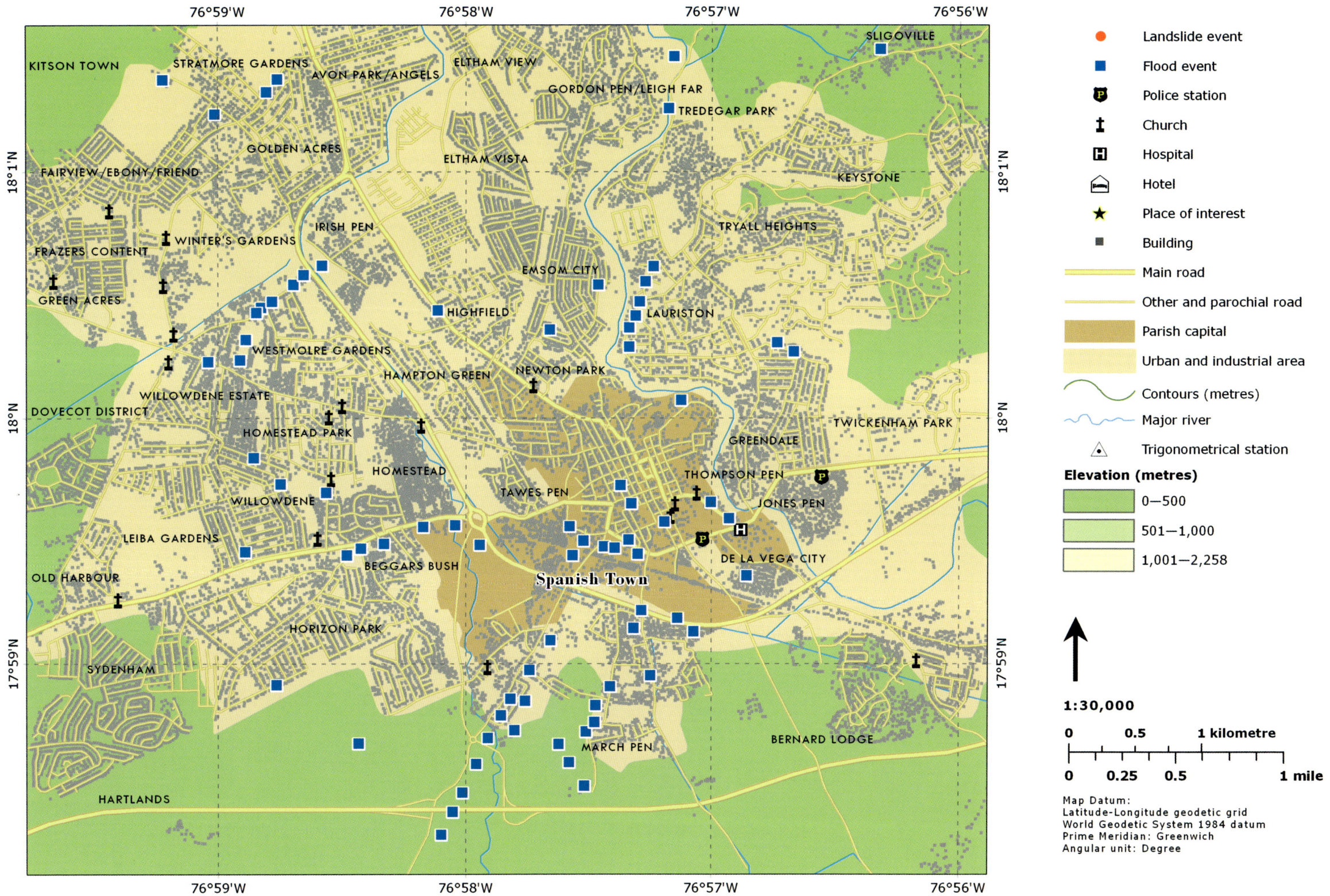

Map 3.14.2 Spanish Town landslide and flood distribution

GLOSSARY

alluvial fan. An outspread, gently sloping mass of alluvium deposited by a stream, especially in an arid or semiarid region where a stream issues from a narrow canyon onto a plain or valley floor. Viewed from above, it has the shape of an open fan, the apex being at the valley mouth (Highland and Bobrowsky 2008).

alluvium. Deposits of clay, silt, sand, gravel, boulders and other particulate matter that has been deposited by a stream or other body of running water in a streambed, on a flood plain, on a delta, or at the base of a mountain (USGS, WS).

aquifer. A geologic rock body or formation that is water bearing and capable of yielding water in sufficient quantity (USGS, WS).

bedrock. A general term for solid rock that lies beneath soil, loose sediment or other unconsolidated material (USGS, WS).

building code. A set of ordinances or regulations and associated standards intended to control aspects of the design, construction, materials, alteration and occupancy of structures that are necessary to ensure human safety and welfare, including resistance to collapse and damage (UNISDR 2009).

climate change. A change in the climate that persists for decades or longer, arising from either natural causes or human activity (UNISDR 2009).

critical facilities. All those elements of the infrastructure that support essential services in a society. They include such things as transport systems, air- and seaports, electricity, water and communications systems, hospitals and health clinics, and fire, police and public administration services (UNISDR 2009).

debris flow (sometimes called mudflow). A flowing mixture of water-saturated debris resembling masses of wet concrete that moves downslope under the force of gravity along water channels. Debris flows are high density and viscosity flows which can carry heavy objects. They consist of material varying in size, from clay particles to rock blocks several tens of metres in maximum dimension, and can travel great distances down valleys at high speeds. Debris flows may climb valley walls on the outsides of bends, and their momentum may also carry them over obstacles. Debris flows confined in narrow valleys or by constrictions in valleys can temporarily thicken and fill valleys to heights of 100 metres or more (USGS, CVO).

digital elevation model. A digital elevation model is a digital file consisting of terrain elevations for ground positions at regularly spaced horizontal intervals (Highland and Bobrowsky 2008).

earthquake. Vibrations in the earth caused by release of energy as a result of rupture and displacement of rock strata across geological faults.

earthquake aftershocks. Smaller earthquakes following a major earthquake.

earthquake epicentre. A point or narrow area on the surface of the earth directly located above the focus.

earthquake focus. The point or narrow area inside the body of the earth where rupture takes place along a fault.

earthquake intensity. A common method for describing size of an earthquake relies on the intensity of ground-shaking obtained from reports on how people felt the shaking and damage done to man-made structures and environment using the Modified Mercalli Intensity Scale (MMI) numbered from I to XII (extreme damage) or the European Macroseismic Scale (EMS) used by the Earthquake Unit in Jamaica. Earthquake events in a close range induce relatively strong shaking and hence higher values on the MMI and EMS scales.

earthquake magnitude. A measure of the size of an earthquake by way of measuring amplitudes of earthquake waves on a seismograph (an instrument that records seismic waves as a function of time) thus quantitatively determining the instrumental magnitude of an earthquake. Commonly expressed magnitudes are Richter; mb, body wave magnitude; ms, surface wave magnitude; Mw moment magnitude, energy released, calculated by taking into consideration the size of the slip on the fault rupture. Large earthquakes result in a larger amplitude of earthquake waves and hence a higher magnitude.

earthquake waves. Energy waves that travel through either the body of elastic rock media as P and S seismic waves or along the surface as surface waves.

fault. A plane of weaknesses across which rocks have lost cohesion and rock strata are displaced.

fault scarp. Generally a steep slope formed by relative displacement of once-adjacent ground surface.

faults in Jamaica. In the Jamaica area, faults have been mapped both on-land and off-shore and have been represented on geological maps. The presence of faults is an indication of geological instabilities. The process of faulting is an important natural geological process that has occurred for tens of millions of years in Jamaica and has shaped the present-day morphology of the island. However, neotectonic faults (that is, faults that have occurred or show displacements during the last 23 million years) are considered to be important when determining potential to generate an earthquake.

flood. An overflow of water onto lands that are used or usable by man and not normally covered by water (USGS, WS).

flood, one-hundred-year. A hundred-year flood does not refer to a flood that occurs once every one hundred years, but to a flood level with a 1 per cent chance of being equalled or exceeded in any given year (USGS, WS).

flood plain. A strip of relatively flat and normally dry land alongside a stream, river or lake that is covered by water during a flood (USGS, WS).

geographic information system (GIS). A computer program and associated databases that permit cartographic information (including geologic information) to be queried by the geographic coordinates of features. Usually the data are organized in "layers" representing different geographic entities such as hydrology, culture, topography and so forth. A GIS permits information from different layers to be easily integrated and analysed (Highland and Bobrowsky 2008).

geologic hazard. A geologic condition, either natural or man-made, that poses a potential danger to life and property. Examples include earthquakes, landslides, flooding, faulting, beach erosion, land subsidence, pollution, waste disposal, and foundation and footing failures (Highland and Bobrowsky 2008).

geologic map. A map on which is recorded the distribution, nature and age relationships of rock units and the occurrence of structural features (Highland and Bobrowsky 2008).

geomorphology. The science that treats the general configuration of the earth's surface; specifically, the study of the classification, description, nature, origin and development of landforms and their relationships to underlying structures, and the history of geologic changes as recorded by these surface features (Highland and Bobrowsky 2008).

ground water. (1) Water that flows or seeps downward and saturates soil or rock, supplying springs and wells. The upper surface of the saturate zone is called the water table.

(2) Water stored underground in rock crevices and in the pores of geologic materials that make up the earth's crust (USGS, WS).

Holocene (geological time). The Holocene epoch is less than ten thousand years before present.

hurricane. A hurricane is a type of tropical cyclone, which is a generic term for a low-pressure system that generally forms in the tropics. The cyclone is accompanied by thunderstorms and, in the Northern Hemisphere, a counterclockwise circulation of winds near the earth's surface. Tropical cyclones are classified as follows:

tropical depression – An organized system of clouds and thunderstorms with a defined surface circulation and maximum sustained winds* of 62 kilometres per hour/38 miles per hour (33 knots**) or less

tropical storm – An organized system of strong thunderstorms with a defined surface circulation and maximum sustained winds of 63–118 kilometres per hour/39–73 miles per hour (34–63 knots)

hurricane – An intense tropical weather system of strong thunderstorms with a well-defined surface circulation and maximum sustained winds of 119 kilometres per hour/74 miles per hour (64 knots) or higher (NOAA/NHC).

*Sustained winds are defined as a 1-minute average wind measured at about 10 metres (33 feet) above the surface.

**1 knot = 1 nautical mile per hour/1.85 kilometres per hour/1.15 statute miles per hour. Abbreviated as "kt".

hydrometeorological hazards. Hydrometeorological hazards include tropical storms, hurricanes, flooding, debris flows, mudflows, droughts and coastal storm surges.

landslide. A landslide is a downslope movement of rock or soil, or both, occurring on the surface of rupture – either curved (rotational slide) or planar (translational slide) rupture – in which much of the material often moves as a coherent or semicoherent mass with little internal deformation (Highland and Bobrowsky 2008).

landslide dam. An earthen dam created when a landslide blocks a stream or river (Highland and Bobrowsky 2008).

land-use planning. The process undertaken by public authorities to identify, evaluate and decide on different options for the use of land, including consideration of long-term economic, social and environmental objectives and the implications for different communities and interest groups, and the subsequent formulation and promulgation of plans that describe the permitted or acceptable uses (UNISDR 2009).

liquefaction. A process where water-saturated, loosely packed, coarse-grained soils lose cohesion and behave in a manner of dense fluids rather than a solid mass especially during the passage of earthquake waves.

lithology. The mineralogical and physical character of rocks.

mitigation. The lessening or limitation of the adverse impacts of natural hazards and related disasters.

mudflow. A general term for a mass-movement landform and process characterized by a flowing mass of predominately fine-grained earth material possessing a high degree of fluidity during movement. The water content may range up to 60 per cent (Highland and Bobrowsky 2008).

natural disaster. A natural hazard event that is severe and extensive, causes widespread human, material and/or environmental losses and a serious disruption of the functioning of a society for a long period of time, and exceeds the ability of affected society to cope using only its own resources.

natural hazard. A potentially damaging physical event, phenomenon or process that can be single, sequential or combined in its origin and effects (for example, earthquakes, hurricanes, landslides, flooding). Natural hazards may cause loss of life or injury, property damage, social and economic disruption, and/or environmental degradation. Each natural hazard is characterized by its location, intensity, frequency and probability.

plate tectonics. A geological model in which the earth's lithosphere is divided into a number of segments or plates that move in relation to other plates along plate boundary zones by sliding past, convergence or divergence.

preparedness. The knowledge and capacities developed by governments, professional response and recovery organizations, and communities and individuals to effectively anticipate, respond to and recover from the impacts of likely, imminent or current hazard events or conditions (UNISDR 2009).

prevention. The outright avoidance of adverse impacts of natural hazards and related disasters.

relief. The difference in elevation between the high and low points of a land surface (Highland and Bobrowsky 2008).

retrofitting. Reinforcement or upgrading of existing structures to become more resistant and resilient to the damaging effects of hazards (UNISDR 2009).

risk. Risk is expressed as the expected value of societal, economic and environmental losses, including deaths, injuries, property damage and so on, that would likely be caused by a natural hazard. It is a function of the hazard and vulnerability.

Saffir-Simpson Hurricane Wind Scale. The Saffir-Simpson Hurricane Wind Scale is a 1 to 5 categorization based on the hurricane's intensity at the indicated time. The scale provides examples of the type of damage and impacts in the United States associated with winds of the indicated intensity (NOAA/NHC). The following table shows the scale broken down by winds:

Category	Wind Speed (kilometres per hour/miles per hour)	Damage
1	119–154/74–95	Very dangerous winds will produce some damage
2	155–177/96–110	Extremely dangerous winds will cause extensive damage
3	178–209/111–130	Devastating damage will occur
4	210–249/131–155	Catastrophic damage will occur
5	>249/>155	Catastrophic damage will occur

A detailed description of the Saffir-Simpson Hurricane Wind Scale, which was revised in early 2010, is available at http://www.nhc.noaa.gov/sshws.shtml.

sedimentary rock. Rock formed of sediment and, specifically, (1) clastic sedimentary rocks – sandstone, shale, conglomerate formed of fragments of other rock transported from their sources and deposited in water, and (2) chemically formed or from organisms, such as most limestone.

storm surge. An abnormal rise in sea level accompanying a hurricane or other intense storm, and whose height is the difference between the observed level of the sea surface and the level that would have occurred in the absence of the cyclone. Storm surge is usually estimated by subtracting the normal or astronomic high tide from the observed storm tide (NOAA/NHC).

tsunami. A long-period ocean wave caused by fault displacement of ocean floor, or submarine landslides, or the collapse of volcanic edifices or the pouring of lava flows in sea water.

volcaniclastic sediments and **volcaniclastic rocks.** Include the entire range of fragmental products deposited directly by explosive or effusive volcanic activity as sedimentary rocks. They may include non-volcanic rock fragments.

vulnerability. The degree of expected damage or loss to a given element at risk (for example, a school building) or set of such elements (for example, a cluster of school) resulting from the occurrence of a natural hazard of a given magnitude and expressed on a scale from 0 (no damage) to 10 (total loss).

weathering. The destructive process by which earth and rock materials exposed to the atmosphere undergo physical disintegration and chemical decomposition resulting in changes in colour, texture, composition or form. Processes may be physical, chemical or biological (Highland and Bobrowsky 2008).

REFERENCES AND FURTHER READING

Ahmad, R. 1986. Recent earth movements at Preston, St Mary, Jamaica. *ODIPERC News* 1 (2): 1–5.

———. 1989a. Earthquake-induced landslides in Jamaica. *Caribbean Landslides Working Group Newsletter* 1 (2): 2–7.

———. 1989b. Geohazards in Jamaica and the Caribbean: The landslide problem. Parts 1 and 2. *UNESCO Courier*, Caribbean supplement 3 (6): 2–4 and 3 (7): 2–4.

———. 1991. Landslides triggered by the rainstorm of May 21–22, 1991. *Jamaican Journal of Science and Technology* 2 (1): 1–13.

———, ed. 1992. *Natural hazards in the Caribbean, special issue Journal of the Geological Society of Jamaica* (12).

———. 1993. Geology of the Blue Mountain Range. In *Blue Mountain guide*, ed. M. Hodges, 19–27. Kingston: Natural History Society of Jamaica.

———. 1995. Landslides in Jamaica: Extent, significance and geological zonation. In *Environment and development in small island states: The Caribbean*, ed. D. Barker and D.F.M. McGregor, 147–69. Kingston: The Press, University of the West Indies.

———. 1996. The Jamaican earthquake of January 13, 1993: Geology and geotechnical aspects. *Journal of Geological Society of Jamaica* 30:15–31.

———, ed. 1997. *Natural hazards and hazard management in the greater Caribbean and Latin America*. Publication no. 3. Kingston: Unit for Disaster Studies, University of the West Indies.

———. 1999a. The Palisadoes and Port Royal, Jamaica: Geological environment, natural hazards and disasters. Post-conference excursion, 14 January. Fourth Conference, Faculty of Pure and Applied Sciences, Unit for Disaster Studies, Department of Geography and Geology, University of the West Indies, Kingston, Jamaica.

———. 1999b. Primer on earthquake hazards and disasters in Jamaica. *Caribbean Geography* 10 (2): 124–36.

———. 2001. Natural hazard maps in Jamaica. *Caribbean Geography* 12 (2): 90–107. [Published 2004.]

———. 2003. Developing early warning systems in Jamaica: Rainfall thresholds for hydrological hazards. Paper presented at the National Disaster Management conference, Ocho Rios, St Ann, Jamaica, 9–10 September.

———. 2004a. *Hazard mapping and site suitability with special reference to Kingston and St Andrew, Jamaica*. Kingston: Unit for Disaster Studies and Department of Physics, University of the West Indies.

———. 2004b. *Hazard maps for landslides and earthquakes, Jamaica*. Kingston: Unit for Disaster Studies and Department of Physics, University of the West Indies.

———. 2004c. *Natural hazards profile of Jamaica*. Kingston: Unit for Disaster Studies and Department of Physics, University of the West Indies.

———. 2004d. *Sediment: Water floods in the Caribbean*. Kingston: Unit for Disaster Studies and Department of Physics, University of the West Indies.

———. 2005a. *Coastal vulnerability and tsunami hazard in Jamaica*. Kingston: Unit for Disaster Studies and Department of Physics, University of the West Indies.

———. 2005b. Living on the edge: Landslide hazard in the Jacks Hill area, Upper St Andrew, Jamaica – Excursion guide. In *Proceedings of the seventh conference of the Faculty of Pure and Applied Sciences, University of the West Indies, Mona*, ed. D. Coore, 17–26. Kingston: Faculty of Pure and Applied Sciences, University of the West Indies.

———. 2007. Risk management, vulnerability and natural disasters in the Caribbean. Report prepared for the International Federation of Red Cross, May. http://www.proventionconsortium.org/themes/default/pdfs/Forum08/Caribbean_Ahmad.pdf

———. 2008. A new examination of floods in the region: Debris floods and debris flows in the Caribbean. In *Enduring Geohazards in the Caribbean*, ed. S.M.J. Baban, 141–56. Kingston: University of the West Indies Press.

Ahmad, R., et al. 1999. *Landslide hazard mitigation and loss- reduction for the Kingston Metropolitan Area, Jamaica*. Publication no. 6. Kingston: Unit for Disaster Studies, University of the West Indies. http://www.oas.org/cdmp/document/kma/udspub7.htm

Ahmad, R., S.M.J. Baban, K. Sant and A. Chinchali. 2004. Flooding and landslides in the West Indies: Digging deeper into the dirt. *Newsletter of the Geological Society of Trinidad and Tobago* (March): 9–13. http://www.gstt.org/publications/The%20Ham merMar20042.pdf

Ahmad, R., B.E. Carby and P.H. Saunders. 1993. The impact of slope movements on a rural community: Lessons from Jamaica. In *Natural disasters: Protecting vulnerable communities*, ed. P.A. Merriman and C.W.A. Browitt, 447–60. London: Thomas Telford.

Ahmad, R., J. Clark, P.A. Manning and M. McDonald. 1997. The potential of bio-engineering in slope stabilization: A case study from Jamaica. In *Natural hazards and hazard management in the greater Caribbean and Latin America*, ed. R. Ahmad, 112–23. Publication no. 3. Kingston: Unit for Disaster Studies, University of the West Indies.

Ahmad, R., J. DeGraff and J.P. McCalpin. 1999. Landslide loss reduction: A guide for the Kingston Metropolitan Area, Jamaica. In *Landslide hazard mitigation and loss-reduction for the Kingston Metropolitan Area, Jamaica*. Publication no. 6. Kingston: Unit for Disaster Studies, University of the West Indies. http://www.oas.org/cdmp/document/kma/udspub6.htm (accessed 20 January 2011).

Ahmad, R., A.H. Earle, P. Hughes, R. Maharaj and E. Robinson. 1993. Landslide damage to the Boar River water supply pipeline, Bromley Hill, Jamaica: Case study of a landslide caused by Hurricane Gilbert. *Bulletin of the International Association of Engineering Geology* 47:59–70.

Ahmad, R., and P. Lyew-Ayee Jr. 2005a. Landslides in north-eastern St Andrew, Jamaica. Rapid assessment, 16 August. Unit for Disaster Studies and Mona Informatix Ltd, University of the West Indies, Jamaica.

———. 2005b. The shaping of Kingston by its urban geology: Field trip 2. In *Programme, abstracts and field guides: The Geological Society of Jamaica fiftieth anniversary conference*, ed. S.F. Mitchell, 26–41. Kingston: Geological Society of Jamaica and the Department of Geography and Geology, University of the West Indies.

Ahmad, R., and M. Mason. 2007. A socioeconomic impact assessment of a 1907-type earthquake on the Jamaican economy (2005). Report of the Disaster Risk Reduction Centre, University of the West Indies, Kingston.

Ahmad, R., and J.P. McCalpin. 1999. *Landslide susceptibility maps for the Kingston Metropolitan Area, Jamaica, with notes on their use.* Publication no. 5. Kingston: Unit for Disaster Studies, University of the West Indies. http://www.oas.org/cdmp/document/kma/udspub5.htm (accessed 20 January 2011).

Ahmad R., D. Miller and D. Rowe. 2002. Landslides related to precipitation in Eastern Jamaica. In *Atlas of probable storm effects in the Caribbean Sea.* [This atlas is available in MS PowerPoint format. It is divided into twenty-one parts, for a total of 18 Mb. Due to its size, it is distributed on CD only (final CDMP CD or from CIMH). http://www.oas.org/CDMP/document/reglstrm/index.htm (accessed 20 January 2011).]

Ahmad, R., and E. Robinson. 1986. Neo-tectonic faults and landslides in Jamaica. Abstracts. Second Caribbean Conference on Natural Hazards and Disasters, Kingston, Jamaica.

———. 1994. Geological evolution of the Liguanea Plain: The landslide connection. In *Proceedings of the first conference, Faculty of Natural Sciences, the University of the West Indies, Mona,* ed. M. Greenfield and R. Robinson, 22–23. Kingston: Faculty of Natural Sciences, University of the West Indies.

Ahmad, R., F.N. Scatena and A. Gupta. 1993. Morphology and sedimentation in Caribbean montane streams: Examples from Jamaica and Puerto Rico. *Journal Sedimentary Geology* 85:157–69.

Brabb, E.E., and B.L. Harrod, eds. 1989. *Landslides: Extent and economic significance.* Rotterdam: Balkema.

Burke, K. 1967. The Yallahs Basin: A sedimentary basin southeast of Kingston, Jamaica. *Marine Geology* 5:45–60.

Burke, K., J. Grippi and A.M.C. Sengor. 1980. Neogene structures in Jamaica and the tectonic style of the Northern Caribbean plate boundary zone. *Journal of Geology* 88:375–86.

Carby, B.E., and R. Ahmad. 1995. Vulnerability of road and water systems to hydro-geological hazards in Jamaica. *Built Environment* 21:145–53.

Chubb, L.J. 1952. A subsidence in the mountains of Jamaica. *Colonial Geology and Mineral Resources* 3:127–32.

Coates, A.G. 1977. Jamaican coral-rudist framework and their geologic setting in reefs and related carbonates: Ecology and sedimentology. *American Association of Petroleum Geologist* 4:83–87.

Cornish, V. 1908. The Jamaica earthquake (1907). *Geographical Journal* 31:245–76.

Costa, J.E. 1988. Rheologic, geomorphic, and sedimentologic differentiation of water floods, hyporconcentrated flows, and debris flows. In *Flood geomorphology*, ed. V.R. Baker, R.C. Kochel and P.C. Patton, 112–22. London: John Wiley and Sons

Dalling, J.W., and S. Irenmonger. 1993. Preliminary estimate of landslide disturbance in the Blue Mountains, Jamaica. *Caribbean Journal of Science* 30:290–92.

Day, M.J. 1982. The influence of some material properties on the development of tropical karst terrain. *Transactions British Cave Research Association* 9:27–37.

DeGraff, J.V., R. Bryce, R.W. Jibson, R. Mora and C.T. Rogers. 1989. Landslides: Their extent and significance in the Caribbean. In *Landslides: Extent and economic significance*, ed. E.E. Brabb and B.L. Harrod, 51–80. Rotterdam: Balkema.

De la Beche, H.T. 1827. Remarks on the geology of Jamaica. *Transactions of the Geological Society of London* 2, ser. 2: 143–94.

Dikau, R., D. Brunsden, D., L. Schrott and M-L. Isben, eds. 1997. *Landslide recognition.* Chichester: John Wiley.

Draper, G. 1990. Jamaica. In *The geology of North America*, vol. H., *The Caribbean Region*, ed. G. Dengo and J.E. Case, 120–27. Boulder: Geological Society of America.

————. 1998. Geological and tectonic evolution of Jamaica. *Contributions to Geology* 3:3–9.

Dumbleton, M.J., G. West and D. Newill. 1966. The mode of formation, mineralogy and properties of some Jamaican soils. *Engineering Geology* 1:235–49.

Earle, A.H. 1991. Landslides in the Rio Minho watershed in central Jamaica. MPhil thesis, University of the West Indies, Jamaica.

Elsner, J.B., and A.B. Kara. 1999. *Hurricanes of the North Atlantic*. New York: Oxford University Press.

Eyre, L.A. 1992. The effects of environmental degradation in the Cane River and Rio Minho Watersheds, Jamaica: A commentary. In *Natural hazards in the Caribbean*, ed. R. Ahmad, special issue *Journal of the Geological Society of Jamaica* (12): 57–65.

Fuller, M.L. 1907. Notes on the Jamaica earthquake. *Journal of Geology* 15:696–721.

Goreau, T., and K. Burke. 1966. Pleistocene and Holocene geology of the island shelf near Kingston, Jamaica. *Marine Geology* 4:207–25.

Green, G.W. 1977. *Structure and stratigraphy of the Wagwater Belt*. No. 48. Kingston: Overseas Geology and Mineral Resources.

Gupta, A. 1975. Stream characteristics in eastern Jamaica: An environment of seasonal flow and large floods. *American Journal of Science* 275:825–47.

Gupta, A., and Ahmad, R. 1999. Urban steeplands in the tropics: An environment of accelerated erosion. *GeoJournal* 49:143–50. [Published in 2000.]

————. 2000. Geomorphology and the urban tropics: Building an interface between research and usage. In *Changing the face of the earth: Engineering geology – Proceedings of the twenty-eighth Binghamton Symposium, 28 August–3 September 1997*, ed. J.R. Giardino, R.A. Marston and M. Morisawa, 133–49. Bologna: Elsevier.

Hall, M. 1907. *Third report on earthquakes in Jamaica, the Great Earthquake of January 14th, 1907 and the after shocks*. Special issue of *Weather Report*, no. 337. Kingston: Government Printing Office.

Harris, N., and M. Rammelaere. 1986. The 1938 Millbank slide and 1940 Chelsea slide, Portland. Special report prepared by Geological Survey Division, Kingston, for the Office of Disaster Preparedness, Government of Jamaica.

Highland, L.M., and P. Bobrowsky. 2008. *The landslide handbook: A guide to understanding landslides*. US Geological Survey circular no. 1325. Reston, VA: US Geological Survey.

Hill, R.T. 1899. The geology and physical geography of Jamaica: Study of a type of Antillean development. *Bulletin of the Museum of Comparative Zoology* 34, ser. 4.

Hill, V.G. 1978. Distribution and potential-clays in Jamaica. *Jamaica Journal* 42:64–75.

Horsfield, W.T. 1973a. Late Tertiary and Quaternary crustal movements in Jamaica. *Journal of the Geological Society of Jamaica* 13:6–13.

————. 1973b. Geological Society of Jamaica field meeting in the Hope River Gorge, Sunday, February 18, 1973, led by W.T. Horsfield. *Journal of the Geological Society of Jamaica* 13:40.

————. 1974. Major faults in Jamaica. *Journal of the Geological Society of Jamaica* 14:1–14.

————. 1975. Quaternary vertical movements in the Greater Antilles. *Bulletin of Geological Society of America* 86:933–38.

Horsfield, W.T., and M.J. Roobol. 1974. A tectonic model for the evolution of Jamaica. *Journal of the Geological Society of Jamaica* 14:31–38.

Halcrow, Sir William, and partners. 1998. Geology and natural hazards. Technical report no. 4, South Coast Sustainable Development Study, prepared for the Government of Jamaica.

Hoek, E., and J. Bray. 1981. *Rock slope engineering*. Rev. 3rd ed. London: Institution of Mining and Metallurgy.

Howell, D.G., E.E. Brabb and R. Ahmad. 1999. Interest in landslide hazard information: Parallels between Kingston, Jamaica and the San Francisco Bay region, USA. In *Landslides: Proceedings of the ninth international conference and field trip on landslides, Bristol*, ed. J.S. Griffiths, M.R. Stokes and R.G. Thomas, 73–79. Rotterdam: Balkema.

Hubbard, R., and J. Fermor. 1972. Landslides on Jamaican roads: An appraisal of causes. Research Notes no. 7, Department of Geography, University of the West Indies.

Hughes, I.G. 1964. Annual report of the Geological Survey Department for the year ended 31st March 1964. Government of Jamaica, Geological Survey Department, Kingston.

Isaacs, M.C. 1985. A brief account of significant twentieth century earthquakes in Jamaica. *Journal Geological Society of Jamaica* 23:25–34.

————. 1987. Seismological investigations in Jamaica: A review. In *Proceedings of a workshop on the status of Jamaican geology: Special issue*, ed. R. Ahmad, 197–224. Kingston: Geological Society of Jamaica.

Jackson, T.A., ed. 1981. June 1979 floods in Jamaica. *Journal Geological Society of Jamaica* 20.

Keefer, D.K. 1984a. Landslides caused by earthquakes. *Geological Society of America Bulletin* 95:406–21.

Kingston Metropolitan Area seismic hazard assessment: Final report. 1999. http://www.oas.org/CDMP/document/kma/seismic/kma1.htm (accessed 20 January 2011).

Lander, J.F. 1997. Caribbean tsunamis: An initial history. In *Natural hazards and hazard management in the greater Caribbean and Latin America*, ed. R. Ahmad, 1–18. Publication no. 3. Kingston: Unit for Disaster Studies, University of the West Indies.

Larsen, M.C., and Parks, J.E. 1997. How wide is a road? The association of roads and mass-wasting in a forested montane environment. *Earth Surface Processes and Landforms* 22:835–48.

Lyell, Charles. 1837. *Principles of geology*. Book 2. 5th ed. London.

Maharaj, R.J. 1992. Geotechnics and zonation of landslides in Upper St Andrew, Jamaica, West Indies. MPhil thesis, University of the West Indies, Jamaica.

Mann, P., and K. Burke. 1984. Neotectonics of the Caribbean. *Reviews of Geophysics and Space Physics* 22 (4): 309–62.

———. 1990. Transverse intra-arc rifting Palaeogene Wagwater belt, Jamaica. *Marine and Petroleum Geology* 7:410–27.

Mann, P., G. Draper and K. Burke. 1985. Neotectonics of a strikeslip restraining bend system, Jamaica. In *Strike-slip deformation, basin formation and sedimentation*, ed. K.T. Biddle and N. Christie-Black, 211–26. Special publication no. 37 of the Society of Economic Palaeontologists and Mineralogists.

Mann, P., C. Schubert and K. Burke. 1990. Review of Caribbean neotectonics. In *The geology of North America*, vol. H., *The Caribbean Region*, ed. G. Dengo and J.E. Case, 307–38. Boulder: Geological Society of America.

Manning, P., T. McCain and R. Ahmad. 1992. Landslides triggered by 1988 hurricane Gilbert along roads in the Above Rock area, Jamaica. In *Natural hazards in the Caribbean*, ed. R. Ahmad, special issue, *Journal of the Geological Society of Jamaica* (12): 34–53.

McFarlane, N.A., comp. 1977. *Geological map of Jamaica, 1:250,000*. Kingston: Ministry of Mining and Natural Resources.

McGregor, D.F.M. 1988. An investigation of soil status and land use on a steeply sloping hillside, Blue Mountains, Jamaica. *Singapore Journal of Tropical Geography* 9:60–71.

———. 1995. Soil erosion, environmental change and development in the Caribbean: A deepening crisis. In *Environment and development in the Caribbean: Geographical perspectives*, ed. D. Baker and D.F.M. McGregor, 189–208. Kingston: The Press, University of the West Indies.

McGregor, D.F.M., and D. Barker. 1991. Land degradation and hillside farming in the Fall River Basin, Jamaica. *Applied Geography* 11:143–56.

Meyerhoff, A.A., and E.A. Krieg. 1977. Petroleum potential of Jamaica. Special report for the Ministry of Mining and Natural Resources, Kingston.

Miller, S., T. Brewer and N. Harris. 2009. Rainfall thresholding and susceptibility assessment of rainfall-induced landslides: Application to landslide management in St Thomas, Jamaica. *Bulletin of Engineering Geology and the Environment* 68:539–50.

Miller S., N. Harris, S. William and S. Bhalai. 2007. Landslide susceptibility assessment for St Thomas, Jamaica, using geographical information system and remote sensing methods. In *Mapping hazardous terrain using remote sensing*, ed. R.M. Teeuw 77–91. Special publication of the Geological Society of London, vol. 283.

Mines and Geology Division Jamaica (MGD). 1974. *Kingston geological sheet 25 with marginal notes, 1:50,000*. Kingston: Ministry of Mining and Natural Resources.

———. 1978a. *Annotto Bay geological sheet 24 with marginal notes, 1:50,000*. Kingston: Ministry of Mining and Natural Resources.

———. 1978b. *Port Maria geological sheet 21 with marginal notes, 1:50,000*. Kingston: Ministry of Mining and Natural Resources.

———. 2004a. *Landslide susceptibility of St Mary, Jamaica, scale 1:50,000*. Kingston: Ministry of Agriculture and Lands.

———. 2004b. Landslide susceptibility map, St Thomas. Unpublished map and guidance notes, Mines and Geology Division, Ministry of Industry and Mining, Kingston, Jamaica.

———. 2007. *Landslide susceptibility of Portland, Jamaica, scale 1:50,0000*. Kingston: Ministry of Agriculture and Lands.

———. 2009. *Landslide susceptibility of St Thomas, Jamaica, scale 1:50,000*. 3rd ed. Kingston: Ministry of Energy and Mining.

———. N.d. *Linstead geological sheet 19 with marginal notes, 1:50,000*. Kingston: Ministry of Mining and Natural Resources.

Ministry of Finance and Planning. 1971. *National Atlas of Jamaica*. Kingston: Town Planning Department, Ministry of Finance and Planning.

Ministry of Public Utilities, Mining and Energy. 1994a. *Geological survey Jamaica*. Blue Mountain geological sheet 13 with explanation of the geology, Jamaica 1:50,000. Geological series (metric ed.). Map comp. M. Rammealeare and E. Parkes, and side margin account by G. Green, D. Holliday and I.G. Balkissoon. Kingston: Geological Survey Division, Ministry of Public Utilities, Mining and Energy.

———. 1994b. *Geological survey Jamaica*. Kingston geological sheet 18 (provisional) with explanation of the geology, Jamaica 1:50,000. Geological series (metric ed.). Map and side margin account comp. M. Rammeleare, E. Parkes, A.R.D. Porter and J.H. Bateson. Kingston: Geological Survey Division, Ministry of Public Utilities, Mining and Energy.

McDonald, F.J., and J. Turnovsky. 1978. Physical development and associated seismic risk in Jamaica. Report for the Mines and Geology Division, Ministry of Mining and Natural Resources, Kingston.

Moore, H.D. 1992. Report on structural map series, metallic minerals survey: Jamaica, phase 2. Geological Survey Division, Ministry of Mining and Commerce of Jamaica, Kingston.

National Environment and Planning Agency (NEPA). 2006. *Annual Report: Financial Year 2005/2006*. Kingston: NEPA.

Natural Disaster Research, Inc., Earthquake Unit, University of the West Indies and Mines and Geology Division, Ministry of Mining and Energy. 1999. Kingston Metropolitan Area seismic hazard assessment final report and appendix. Prepared for the US Agency for International Development/Organization of American States, Caribbean Disaster Mitigation Project.

Natural Resources Conservation Authority (NRCA). 1992. Jamaica national report for UNCED, Brazil, June. NRCA, Kingston.

Naughton, P.A. 1976. The assessment of natural hazard risk as a basic tool in environmental land use management in the Kingston Metropolitan Area, Jamaica, West Indies. PhD thesis, University of the West Indies, Jamaica.

———. 1984. Flood and landslide damage–repair cost correlations for Kingston, Jamaica. *Caribbean Geography* 1:198–202.

Northmore, K.J., R. Ahmad, E. O'Connor, D. Greenbaum, A.J.W. McDonald, C.J. Jordan, A.P. Merchant and S.H. Marsh. 2000. Landslide hazard mapping: Jamaica case study, National Environment Research Council. British Geological Survey technical report WC/00/10, DFID project no. R6839. http://www.bgs.ac.uk/dfid-kar-geoscience/database/reports/bw/WC00010_BW.pdf

O'Hara, M. 1980. Case studies of physical damage caused by the 12th June 1979 flood rains in western Jamaica. In *Proceedings of the ninth Caribbean Geological Conference, Santo Domingo*, 563–74.

———. 1987. Physical hazard and risk assessment in Jamaica, West Indies. In *Planning and engineering geology*, ed. M.G. Cushlaw, F.G. Bell, J.C. Cripps and M. O'Hara, 311–22. *Engineering Geology*. Special publication of the Geological Society, no. 4.

O'Hara, M., and R. Bryce. 1983. A geotechnical classification of Jamaican rocks. Bulletin no. 10, Geological Survey Division, Government of Jamaica.

Organization of American States. 1999. Storm surge hazard mapping for Montego Bay, Jamaica. http://www.oas.org/CDMP/document/kma/mobay/mobay.htm (accessed 20 January 2011).

Paterson, Grant & Watson Ltd. 1992. Jamaica gravity and aeromagnetic map series. Canadian International Development Agency project no. 504/12713. Kingston: Geological Survey Division, Ministry of Production, Mining and Commerce.

Pereira, J. 1977. An engineering seismology study of Jamaica. MSc thesis, Imperial College, London.

Persson C. 1984a. Tropical cyclone: Preliminary hazard assessment, ref. GSD no. D 126, scale 1:250,000. Geological Survey Division, Government of Jamaica.

———. 1984b. Earthquake: Preliminary hazard assessment, scale 1:250,000. Geological Survey Division, Government of Jamaica.

Robinson, E. 1963. Quaternary deposits. In *Synopsis of the geology of Jamaica: An explanation of the 1958 Provisional Geological Map of Jamaica*, ed. V.A. Zans, L.J. Chubb, H.R. Versey, J.B. Williams, E. Robinson and D.L. Cooke, 50–54. Bulletin no. 4. Kingston: Geological Survey Department.

———. 1975. Jamaica and offshore banks: Preliminary magnetic anomalies and bathymetry, scale 1:465,000. Mines and Geology Division, Ministry of Mining and Natural Resources, Jamaica.

———. 1994. Jamaica. In *Caribbean geology: An introduction*, ed. S.K. Donovan and T.A. Jackson, 111–28. Kingston: University of the West Indies Publishers Association.

Robinson, E., R. Ahmad, C. Phillip-Jordan and M. Armstrong. 1996. The Burlington landslide, mouth of Rio Grande, Jamaica: An example of an ancient landslide dam? *Journal of the Geological Society of Jamaica* 31:37–42.

Robinson, E., H.R. Versey and J.B. Williams. 1960. The Jamaica earthquake of March 1, 1957. In *Transactions of the second Caribbean Geological Conference 1959, Puerto Rico*, 50–57.

Rural Physical Planning Unit. 1985. *General soil map of Jamaica, 1:250,000*. Mandeville, Jamaica: Rural Physical Planning Unit, Ministry of Agriculture, Central Region.

Sawkins, J.G. 1869. *Report on the geology of Jamaica*. London: Longmans, Green and Co.

Shepherd, J.B. 1971. A study of earthquake risk in Jamaica and its influence on physical development planning. Report for the Ministry of Finance and Planning, Kingston.

———. 1989. Earthquake and volcanic hazard assessment and monitoring in the Commonwealth Caribbean: Current status and needs for the future. In *Proceedings of the meeting of experts on hazard mapping in the Caribbean, 30 November to 4 December, 1987, Kingston, Jamaica*, ed. D. Barker, 50–61.

Shepherd, J.B., and W.P. Aspinall. 1980. Seismicity and seismic intensities in Jamaica, West Indies: A problem in risk assessment. *Earthquake Engineering and Structural Dynamics* 8:315–35.

Simpson, R.H., and H. Riehl. 1981. *The Hurricane and Its Impact*. Baton Rouge: Louisiana State University Press.

Sloane, H. 1809. A letter from Hans Sloane, M.D. and F.R.S. with several Accounts of the Earthquakes in Peru, Oct. 20, 1687; and at Jamaica, Feb. 19, 1683–94; and June 7, 1692. *Philosophical Transactions of the Royal Society of London* 3:626.

Taber, S. 1920. Jamaica earthquakes and the Bartlett trough. *Bulletin of the Seismological Society of America* 10:55–89.

Thomas, M.F. 1996. *Geomorphology in the tropics*. Chichester: John Wiley and Sons.

Tomblin, J.F. 1976. Earthquake risk in Jamaica. *Journal of the Geological Society of Jamaica* 15:16–23.

Tomblin, J.M., and G.R. Robson. 1977. *A catalogue of felt earthquakes for Jamaica, with references to other islands in the Greater Antilles, 1564–1971*. Special publication no. 2. Kingston: Mines and Geology Division, Government of Jamaica.

Turner, A.K., and R.L. Schuster, eds. 1996. *Landslides: Investigation and mitigation*. Transportation Research Board, special report no. 247. Washington, DC: National Academy Press.

Van Dusen, S.R., and D.I. Doser. 2000. Faulting processes of historic (1917–1962) M > 6.0 earthquakes along the north-central Caribbean margin. *Pure and Applied Geophysics* 157:719–36.

Varnes, D.J., ed. 1984. Landslide hazard zonation: A review of principles and practices. UNESCO Natural Hazard Series, no. 3. Paris: UNESCO.

Vincenz, S.A. 1959. Some observations on gamma radiation emitted by a mineral spring in Jamaica. *Geophysical Prospecting* 7:422–34.

Wadge, G., and T.H. Dixon. 1984. A geological interpretation of SEASAT–SAR imagery of Jamaica. *Journal of Geology* 92:561–81.

Water Resources Authority (WRA). 2011. *Water Resources of Jamaica: Fact Book*. Kingston: WRA.

Wiggins-Grandison, M. 1993. The earthquake of January 13, 1993, and implications for earthquake hazard in eastern Jamaica. In *Proceedings of the Caribbean Conference on Natural Hazards: Volcanoes, earthquakes, windstorms, floods, Port of Spain, Trinidad, October 1993*, ed. W.B. Ambeh, 65–76. St Augustine, Trinidad: Seismic Research Unit, University of the West Indies.

———. 1996. Seismology of the January 1993 earthquake. *Journal Geological Society of Jamaica* 30:1–14.

Wiggins-Grandison, M.D., and K. Atakan. 2005. Seismotectonics of Jamaica. *Geophysical Journal International* 160:573–80.

Wiggins-Grandison, M.D., and J. Havskov. 2004. Crustal attenuation for Jamaica, West Indies. *Journal of Seismology* 8:193–209.

Wiggins-Grandison, M.D., and W. Reid. 1993. Seismic hazard in Jamaica. In *The practice of earthquake hazard assessment*, ed. R.K. McGuire, 165–68. Denver: International Association of Seismology and Physics of the Earth's Interior.

Williams, J.B. 1965. Geology and engineering in Jamaica. *Bulletin of the Scientific Research Council of Jamaica* 6:29–45.

Wood, P.A. 1976. The evolution of drainage in the Kingston area. *Journal of the Geological Society of Jamaica* 15:1–6.

Zans, V.A. 1959. Judgment Cliff landslide in the Yallahs Valley. *Geonotes: Journal of Geological Society of Jamaica* 2:43–48.

Zans, V.A., L.J. Chubb, H.R. Versey, J.B. Williams, E. Robinson and D.L. Cooke. 1962. Synopsis of the geology of Jamaica. Bulletin no. 4, Geological Survey Department, Kingston, Jamaica.

ABOUT THE AUTHORS

Parris Lyew-Ayee Jr is the Director of the Mona GeoInformatics Institute at the University of the West Indies, which functions as the geographic information systems (GIS) hub of the Mona campus. He is responsible for overseeing all commercial and research and development activities at the operation. He also lectures final-year students in GIS at the Department of Geography and Geology, as well as several MSc courses offered at the university.

He sits on over a dozen boards and committees, including serving as the Chairman of the Water Resources Authority. His active research includes natural hazards analyses, crime modelling, Martian rock breakdown analyses, transport systems modelling, karst geomorphology, terrain signature diagnoses, geocomputational modelling, geospatial planning for businesses, GPS systems and web mapping.

He is the author or co-author of over thirty peer-reviewed books, papers, book chapters, magazine articles and consultancy technical reports, and is the recipient of numerous awards, grants and commendations at both local and international levels.

Rafi Ahmad is Head of the Unit for Disaster Studies, a Fellow of the Mona GeoInformatics Institute, and lecturer in geology at the Department of Geography and Geology. He has over forty years' experience in the field of geology and natural hazards, nearly thirty of which have been served in Jamaica. He has previously served as head of the Department of Geology, refereed papers for international journals, and served as editor of the *Journal of the Geological Society of Jamaica*. He is on the editorial board of the *Quarterly Journal of Engineering Geology*.

He is an internationally recognized geohazards expert, delivering presentations and courses in numerous countries and territories, and serves or has served as a consultant to many national and international committees and associations. He is a Fellow of the Geological Society of London, as well as member of the Geological Society of Jamaica and the Jamaica Institute of Environmental Professionals.

His research interests include urban geology and geomorphology, natural hazards, neotectonics, remote sensing and aerial photographic interpretation, structural mapping, and engineering geology. He is the author of over 190 peer-reviewed articles, book chapters, articles and technical reports, and is the recipient of numerous local and international awards.